Climate Change as Societal Risk

Mikael Granberg · Leigh Glover

Climate Change as Societal Risk

Revealing Threats, Reshaping Values

Mikael Granberg
The Centre for Societal Risk Research
Karlstad University
Karlstad, Sweden

Leigh Glover
The Centre for Societal Risk Research
Karlstad University
Karlstad, Sweden

ISBN 978-3-031-43960-5   ISBN 978-3-031-43961-2   (eBook)
https://doi.org/10.1007/978-3-031-43961-2

Cover illustration: © Melisa Hasan

This Palgrave Macmillan imprint is published by the registered company Springer Nature
Switzerland AG
The registered company address is: Gewerbestrasse 11, 6330 Cham, Switzerland

Paper in this product is recyclable.

PREFACE

This book is an output of the research program, 'Resilience in Sweden: Governing, Social Networks and Learning' (MSB/2016-6855), financed by the research fund of the Swedish Civil Contingency Agency. It resulted from a long-term collaboration between Mikael Granberg and Leigh Glover that started in 2009 at the Australasian Centre for the Governance & Management of Urban Transport (GAMUT), the University of Melbourne, Australia. The collaboration has produced a number of research articles, research funding proposals, conference presentations, plans, ideas and the book, *The Politics of Adapting to Climate Change* (2020), published by Palgrave Macmillan. The collaboration revolves around issues connected to societal aspects of climate change, such as its impacts on society, climate action by societal actors, environmental/ climate justice and equity and the interdependence between societal development and climate change vulnerabilities. The whole collaboration has been informed by an awareness of the political aspects of society´s effort to address (or not to address) climate change hazards, risks and related impacts. The studies have included climate issues in Australia, Sweden and the wider world.

We thank the Centre for Societal Risk Research (CSR) at Karlstad University, and the Centre for Natural Hazards and Disaster Science (CNDS), Uppsala University (both located in Sweden), for their support in working on this book. Dr. Johanna Gustavsson, Director of CSR and Emilie Hindersson, Coordinator of CSR, are thanked for their support of this volume and the project that underpins it. Our colleagues, Prof. Lars Nyberg and Andreas Pettersson (both of CSR) have also contributed to our efforts. We also like to thank A/Prof. Susie Moloney for the support in putting this book together. Finally, we want to thank our families for putting up with us during the writing of this book.

Input from the (anonymous) reviewer of the book proposal was appreciated and the final manuscript reflects his/her insights and constructive comments, questions and suggestions. We also thank the good people at Palgrave Macmillan for their work in developing our ideas and arguments and finalising this project. Responsibility for the ideas and positions put forth in this book rests, however, firmly with the authors.

We were sorry to learn of the death of Emeritus Prof. Will Steffen of ANU, Canberra, during the writing of this book; we value his contribution and his inspiration to many in this field and we mourn the loss.

Last but not least we want to thank our families for their continuous support!

Örebro, Sweden   Mikael Granberg
May 2023   Leigh Glover

CONTENTS

LIST OF BOXES

LIST OF TABLES

PART I

Considering the Society, Climate and Risk Problem

CHAPTER 1

Introducing Climate Change as Societal Risk

Abstract This chapter introduces the era of climate change in which we are living and connects this to the concept of the Anthropocene. A primer to the concept of risk precedes a discussion of climate change risks and the challenges these entail for society. Central concepts of climate change risks are defined and explained. Societal risk is introduced and contrasted with other risk types used in the climate change discourse. Differences between social risk and societal risk are elucidated. How societal risk can be understood in the climate change context is described. A section about the entire volume, including a statement of its aims, closes the chapter.

Keywords Anthropocene · Climate change · Climate–society nexus · Resilience · Risk · Societal risk · Society · Transformation · Values

CLIMATE CHANGE AND THE ANTHROPOCENE

Ours is the era of climate change of our own making. Climate change's harmful effects cannot be undone and, mostly, will inexorably be borne by our descendants and the future natural world as altered by human hand (IPCC, 2023). Furthermore, the greatest harms will be experienced by the world's poor (IFRC, 2020), a condition evoking the social dimension of risk as "...the risks of climate change to social systems is

© The Author(s), under exclusive license to Springer Nature
Switzerland AG 2023
M. Granberg and L. Glover, *Climate Change as Societal Risk*,
https://doi.org/10.1007/978-3-031-43961-2_1

as much about the characteristics of those systems as it is about changes in environmental systems" (Barnett & Adger, 2007, p. 641). Therefore, climate change is possibly the most profound and transformative challenge to economic, social and political systems that humanity has ever faced. This condition highlights the interdependent relationship constituting the climate–society nexus.

These climate change risks, hazards and vulnerabilities are a function of societies' development over the centuries, especially the model of technological development under the modern industrial state that has given humanity the capacity to alter the global climate (Glover, 2006, p. 1). This new geological epoch, distinguished by the contemporary relationship between social organisation, scientific and technological advances, and exploiting natural resources and ecosystem services at the global scale, is labelled as the Anthropocene (cf., Crutzen & Stoermer, 2000; Steffen et al., 2018). In the Anthropocene, human activities are driving environmental, ecological and even geological change, thereby altering the preconditions for life on this planet, and resulting in new, previously unknown, challenges, problems, hazards and risks. This also fuels calls for transformative societal change, of its functions, its infrastructures (built, informational and social) and its institutions (Levin et al., 2022; Schlosberg et al., 2017). As "…humans are a force of nature in the geological sense" (Chakrabarty, 2009, p. 207), human and natural history are intertwined and drawing a distinction between them is becoming more arbitrary and less meaningful.

Risk: A Primer

Risks are complex, in all meanings of the term, consisting of an interdependent relationship between *exposure* to hazards, *vulnerability* as the potential for loss and *capacity* for effective action that determines the extent of harm or loss (see Fig. 1.1).

Risk relates to the functionality and development of society, technological innovation and people's lives and lifestyles and is understood, formulated and responded to, in widely differing ways within society and between scholarly disciplines. Efforts at recognising, understanding, knowing and interpreting risk are arguably exercises in attempting to grasp the unknown and indeed, to a certain extent, to know the unknowable. And yet, as individuals, as social groups, as institutions, as governmental bodies and as enterprises, we need to know the risks we face as

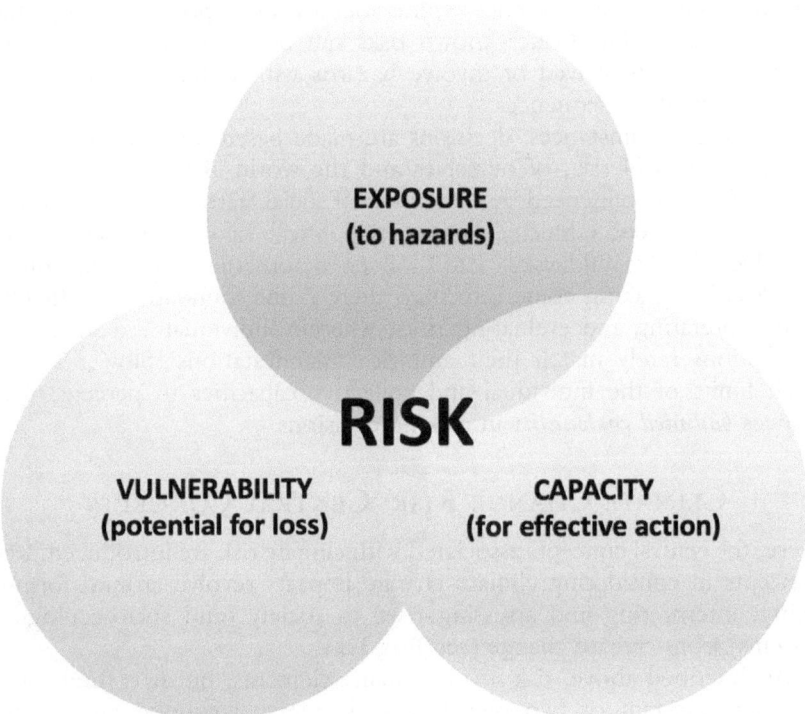

Fig. 1.1   Risk as a combination of exposure, vulnerability and capacity

best we can. Such knowledge is needed to make individual and collective decisions as to how best to respond (or not) to perceived threats to our interests, plans, resources, values and welfare (cf., Kasperson & Kasperson, 2005; Pidgeon et al., 2003).

As individuals and as societies, a key conundrum of risk arises from knowledge boundaries, whereby the entirety of the character of risk cannot be known. Society is perpetually evolving, with accumulating learning, research and technological developments all contributing to the tasks of risk awareness and mitigation. Despite this, our knowledge of risk is dynamic and uncertain, framed by language and context, incompletable and open-ended (Taleb, 2008). Some risks go undetected by risk analysis models and predictions, only becoming manifest at the point of system failure, when societies' vulnerabilities are exposed in ways that can be catastrophic. Other known risks can be of an entirely different character than perceived or involve hazards with unforeseen social and environmental consequences.

In these circumstances, decisions are made based on perceptions and interpretations of risk, of ourselves and the world as we know it. Risk perceptions are influenced by all manner of social factors, notably culture, educational levels, ethnicity, language, political views and social status (cf., Douglas & Wildavsky, 1983). Even before the matters of collective decision making come into play, there is the foundational difficulty in interpretating and evaluating risks, wherein individual and social risk perceptions rarely match their empirical manifestations (Slovic, 2000). Such limits of the individual and collective capacities to perceive risk, evinces *bounded rationality* in making decisions.

CLIMATE CHANGE RISK: CENTRAL CONCEPTS

Here, the central concepts associated with climate risk are introduced. Key concepts in considering climate change impacts revolve around formulating, interpreting and assessing risks to society (and socio-ecological systems) from climate change (see Box 1.1).

As described above, risk involves many elements, but it is fundamentally an amalgam of two constituent elements: likelihood and consequences (Lupton, 1999). When discussing climate change risks, low probability/high consequence events are of particular interest, given the necessity of individuals, corporations, institutions, governments and society to avoid catastrophes. Yet the rarity of the catastrophic limits the opportunities for experience, learning, preparation, modelling and other responses that form part of disaster and risk preparation, mitigation and management (for an overview of responses, see Lopéz-Carresi et al., 2014). We know climate change is happening, there is uncertainty on exactly what will happen, but we know it

> ...will be bad, but we do not know exactly *how* bad it will be *when* and *for whom*. We are therefore in a position where we must decide what to do about the risks climate change threatens in the face of a range of uncertainties. (Hartzell-Nichols, 2017, p. xvii, italics in original)

Box 1.1   Key concepts in climate risk

Adaptation: Responses in social and socio-ecological systems, to ameliorate the costs/ disbenefits of potential or actual hazards and/or to take advantage of opportunities related to the hazard response. In natural systems, these are responses to climate (and related) changes

Cascading Risk: Where a risk in one system creates an additional risk in a related system, promoting a 'chain reaction' effect

Collapse: A dramatic and irreversible change in a system or phenomena, usually multi-causal and involving triggering events and negative feedbacks

Compounding Risk: Risks arising from multiple hazards interacting to create greater hazards (i.e., greater than the sum of the parts); these can be coincidental or occur sequentially (i.e., in a series of occurrences)

Exposure: That extent of openness to risks

Hazards: Those (potential or perceived) attributes/events/occurrences that threaten to damage, degrade or harm those attributes, functions, resources, services and structures that are vulnerable and valued

Resilience: A capacity to respond to hazards that protects the attributes, functions, structures and or values threatened by climate and climate change; resilience can be achieved through adaptations to climate change

Risk: That potential for adverse consequences for human and/or ecological systems

Societal transformation: Involves essential, multifaceted and complex changes to a system that alters elements of its forms, functions, meanings, relationships and/or structures

Tipping Point: A threshold change in a system, usually characterised by abruptness and irreversibility

Vulnerability: That propensity to be harmed, suffer loss or undergo change from the climate system; often assessed as the sensitivity to change/harm or extent of resilience

---

Sources After, Reisinger et al. (2020) and IPCC (2022)

This formulation applies to the known or the 'known unknown'; beyond this boundary lies those risks of the 'unknown unknowns' type, the understanding of which only occurs in hindsight) where *"...what you don't know"* is "...far more relevant than what you do know" (Taleb, 2008, p. xxiii, italics in original). Therefore, decisions must be made, and associated climate risk-reducing measures implemented, under considerable uncertainty. Entering such unknown terrain calls for precaution.

Climate change risks are complex, as they arise through interactions among coupled systems of climate, ecosystems and society (IPCC, 2023). The Intergovernmental Panel on Climate Change (IPCC, the UN climate science advisory body) define risk related to climate change as "...the potential for adverse consequences for human or ecological systems, recognising the diversity of values and objectives associated with such systems" (IPCC, 2022, p. 4). This focus on values highlights risk's social and cultural dimensions.

Over time, our recognition of climate hazards has greatly broadened. One way of considering actual and potential impacts is by recognising the different ways in which they can be presented:

- As weather and climatic extremes,
- As change in the duration, frequency and variability of climatic attributes,
- As ongoing and relatively slow change,
- As cascading/compounding risks occurring in related systems, and
- As unexpected change.

Hazards can be potentially catastrophic and irreversible, such as ice sheet collapses in Greenland and the West Antarctic, Northern permafrost thawing and the collapse of the Labrador Sea Current (Lenton et al., 2019). In the societal realm, the most extreme expression of such change is *societal collapse* that can result from interactions with climate change and other factors. While some hazards are apparent, obvious and essentially 'regular', such as droughts, floods and storms, others are covert, subtle and extremely difficult to observe or predict. In addition, the problem of cascading and compounding risks means that activities, communities and locations distant in space and time from one climatic hazard can be exposed to an unexpected hazard (Lawrence et al., 2020).

Vulnerability to weather-related hazards are highly socially differentiated (Malloy & Ashcraft, 2020). A relatively minor hazard in one community can cause profound social harm and yet another community can experience a major hazard and emerge relatively unscathed. Economic, geographic and social factors profoundly influence social resilience and vulnerability to hazards and, as discussed below, similarly influence climate adaptations and disaster recovery. In short, the vulnerability of a system to a hazard depends not only on hazard exposure and the consequences of exposure, but also on the capacity of the system and any adaptive measures taken.

When it comes to the social and ecological risks of climate change, social factors influence the risks of climate hazards, centring on the vulnerability in the social realm to a hazard, its resilience to actual or potential impacts, the influence and effectiveness of adaptation measures and practices, and the opportunities for transformation leading to reforms that may ultimately reduce vulnerabilities and contribute to other social goals. It is the combination of these factors that, in effect, determines the climate change risks to society.

Resilience is one of those somewhat rare descriptors, where a concept has similar meaning whether it is applied to individuals, collectives or societies (see, e.g., Alexander, 2013). Climate resilience in the social realm is the capacity for an individual, group, corporation or institution to successfully recover from a disturbance or disruption of climatic origin. In systemic terms, resilience is the attribute of recovery without permanent loss of system structure or function (GCA, 2019) and has become one of the central themes in national, state and urban climate adaptation responses (Meerow et al., 2019). Resilience in socio-ecological realms is dynamic and determined by context (i.e., the characteristics of the threat and of what is threatened). Implicit in climate resilience is an awareness of the risks of climate impacts and that this is a malleable capacity that can be managed and enhanced. Vulnerability assessment implies evaluating existing resilience to the threats of current or forecast climate change impacts; vulnerability and resilience are seemingly two sides of the same coin (Sovacool, 2018).

However, as Mark Pelling (2011) argues, social resilience can also be detrimental in the face of change, as it implies returning to starting conditions, whereby recovery leaves the system re-exposed to future disturbances. Properly reducing vulnerability to climate impacts, by definition, involves adopting transitional or transformative responses, whereby the scale and scope of future risks are reduced by changing the system's structure and function.

In considering the climate–society nexus in terms of the relationship between climatic change and climate-exposed social systems, given the variability in minor climatic and related systems, there is already a considerable degree to which routine social and economic behaviour can continue, i.e., social systems have sufficient flexibility within a 'coping' capacity. Spontaneous adaptations to climate change can occur without deliberate intervention, such as market adjustments in economic systems. Sloat et al. (2020), for example, describe the migration of rainfed grain cropping globally over the period 1973–2012 in response to climate changes.

Adaptation measures, through social actions, are taken by responding to current impacts or anticipating forecast impacts. Decision-making by risk managers rests on the fundamental reckoning of social and natural systems' *climate sensitivity*, the social value at risk and risk management's *adaptive capacities* (Eckstein et al., 2020). In these decisions, the uncertainty over the likelihood of change and the uncertainty over the extent, magnitude and scale of the change in question must be acknowledged in some way. The extent to which losses occur incrementally through exposure to climate change defines the *vulnerability* to losses, in conjunction with relevant non-climatic variables.

Adaptation occurs as a response to change or predicted change and might reduce consequences of said change in social and natural systems; it can be anticipatory, autonomous or planned. Society's adaptation to climate change impacts, either in response to, or in anticipation of, changes can involve all or some of the ensuing sequential activities: Information gathering, interpretation and knowledge formation, decision-making institutions and processes, and subsequent policy formulation, planning and implementation actions, followed by monitoring and evaluation (for an overview of activities, see, Glover & Granberg, 2020). Furthermore, another societal response is deliberate inaction, i.e., the 'no response option' (Nicholls et al., 1995).

Deliberate societal adaptations involve assessments of both climatic change impacts and of the relevant social systems (Leichenko & O'Brien, 2019). Realising or anticipating change beyond existing social capacities evokes the need for adaptation, as social interests are vulnerable to loss beyond this level of system change. In most instances, the extent of climate change that exceeds coping capacity will be understood as a threshold event. From a social science perspective, social thresholds might be of an arbitrary character set by groups, firms, governments or individuals.

Somewhat neglected in the discourse over these risks is the consequence of uncertainty over social thresholds (Adger et al., 2009). Adaptation measures designed in response to forecast climate change impacts can be inadequate or overwhelmed if the realisation of the impacts exceeds forecasts. In other words, adaptations are responses to forecast risks that have the effect of raising the threshold at which unacceptable risks occur, but in most cases cannot be assumed as providing absolute security. *Climate-proofing*, therefore, should not be taken as an absolute condition; in nearly all cases, it is only relative to the accuracy and confidence of the risk forecast of the impacts in question (Fankhauser & Schmidt-Traub, 2011).

Resilience engages choices over preservation, which entails concern over the values at risk of loss. A contrary perspective considers the implications of *societal transformation*, built on reforming social systems to reconfigure risks and seek associated social gains. Transformation can be strategically chosen by societal actors or forced on society by climate and/or social realities and these two types of drivers have radically different impacts and potential outcomes for society (Winkelmann et al., 2022). Climate change, therefore, can be perceived as providing a 'critical juncture' where path-dependent institutions, policies and practices have an opening to change pathways (this does not mean, however, that new pathways necessarily will develop). From this perspective, risks are compounded, interconnected and multi-dimensional because risks and society's socio-political and technological responses to them interact (Assmuth et al., 2010).

These contrasting pathways—adaptation, resilience, transition and transformation—have several implications for societal risk. Political values and implications arise from public policy activities. For example, Vale (2014) suggests that resilience can only be a useful concept if it is part of efforts to address social disadvantage. Meerow et al. (2019, p. 794) points out that community resilience is highly unequal and that it is unclear "...who is truly benefiting from..." the resilience efforts of public actors. Issues of why and how 'goods and bads' are distributed within society through climate responses involves social decisions on the identity of what/whom is threatened, the identity of those threats and the beneficiaries of any responses.

Introducing Societal Risk

As a global society, our concerns over risk are those of loss or harm to that which we value, either in their extant or potential forms (Hartzell-Nichols, 2017). As individuals, and as societies, we value so much; the benefits of our lifeworld are a complex construction of a multitude of elements from the natural and social worlds, very little of which we are willing to let go, or be deprived of, without feeling a material, conceptual or spiritual loss and a diminishing of our individual lives and collective well-being, in some respect. For a social scientist to consider societal risk, is to engage in some of the most integral and conflicted issues in the social sciences. In the public realm, risk generates some of the deepest public debates and controversies, for no other reason that it involves political machinations over the hazard exposure, distributing or managing disbenefits and allocating public resources and opportunities.

Much could be written over the identity of society, and indeed much has and the resulting ambiguity and differences pose many conceptual barriers for scholars dealing with societal risk, such that the majority seemingly prefer to skip over the matter of defining 'society' entirely. Rather than formulate an original social theory (an overwhelming task), or adopt a singular and accepted conceptual sociological explanation, a more basic and flexible approach to defining society is proposed, drawing on familiar sociological themes. Society is understood as possessing the following identifying features:

- A distinct set of economic, political and social functions,
- Management of socio-ecological systems that furnish resources and services,
- Governance and associated institutions for managing collective interests,
- A common ethos, covering distinctive values and practices, and
- A set of symbolic and material components that comprise a culture.

Additional to these features of belief, control, function, order, purpose, structure and survival is that of dynamism; a central feature of society is change, conflict, development, difference and evolution. It is held that these attributes are universal characteristics of all societies (such as covering hunter-gatherer, horticultural, pastoral, agrarian, industrial and post-industrial types). Further, the anthropological/geographical perspective that societies exist in specific locations provides an identifying feature linking many of these social components, so that a society acquires its distinguishing features from inhabiting a particular place. Recognising that societies are situated in-place allows for multiple scales of society according to the scale at which the place is viewed. Therefore, we do not perceive society as one comprehensive unit (i.e., a single, fixed entity) rather, the concept can be applied at many scales, such as at localities, regions and nations.

We can also be clear of that which does not qualify as a society. An aggregation of individuals, such as a crowd, is not considered to be a society in the way that the concept is understood and applied to describe a collective, identifiable and ordered identity. As indicated above, a society can include businesses, corporations, governments and institutions, but such entities in isolation do not constitute a society. Society takes many forms and expressions and includes an array of arrangements, constellations of groups and institutions in business, civil society and government, and there are many terms describing the social entities threatened by hazards.

Although climatic hazards, society and risk can be rendered into relatively definable forms, societal responses to risk are a considerably more fluid and conditional realm of activity. As a rule, individuals and society do not usually make singular decisions over risk per se, but make constant decisions taking risk into account together with, and in relation to, other factors. These risk-related decisions can readily change over time. Risk is at the tail-end of the constellation of social values and political processes. As many scholars have pointed out, risk cannot be understood as being purely objective because investigating any risk is initiated by an awareness of a hazard(s) (Tulloch, 2008). Risk, therefore, is highly conditional on the circumstances in which it is perceived to exist. It is impossible to make sense of the meaning of risk as the chances of loss realisation without knowing what constitutes this risk and of the consequences of it being realised.

One aspect of macro-scale risks, that receives substantial inquiry and scholarship, concerns the financial system (Campiglio et al., 2018). This constellation of commercial and financial risk can be treated as a distinctive genre, distinct from natural hazard and technological risk assessment and management. Another field of risk studies focusing on the individual is psychology, which identifies a range of interests and values influencing climate risk attitudes, perceptions and response behaviours and the ways in which these differ from that of the ideal established by scientific-technical expectations (Clayton & Manning, 2018; van Valkengoed & Steg, 2019).

Scholarship into climate change risks in the social sphere correspondingly covers a broad array of issues and gives rise to an assemblage of different risk interpretations in the social sphere as to what constitutes a social and societal risk. For clarification, the major risk types need to be differentiated (as summarised in Box 1.2).

Ecological risks usually refer to hazards to the natural world, i.e., meaning the non-human, such as ecosystems, habitats, species and organisms. Although 'environmental risks' is sometimes used for the same

Box 1.2   Contrasting risk types pertaining to climate change

---

Ecological risk: That potential for adverse/harmful outcomes and consequences for ecological entities, processes, systems and values
Social Risk: That potential for adverse/harmful outcomes and consequences for people that need not be interrelated, and for collective groups, including businesses, institutions, governments and non-government organisations. This may include socio-ecological systems. It is often applied in a general sense to distinguish between individual and collective risks
Societal Risk: That potential for adverse/harmful outcomes and consequences for an identified society. This includes socio-ecological systems and may incorporate ecological risks
Systemic Risk: That potential for adverse/harmful outcomes and consequences for society when it is conceptualised as a system, either informally or formally, i.e., an understanding of society based on a general system theory framework (aka, socio-cybernetics). It can refer to the potential for existential loss, as in a system failure, such as might occur to a business, corporation, institution, market or economic sector

---

purpose, that term generally refers to hazards to people mediated through the environment, and usually involves contaminants and pollutants. Decisions over ecological and environmental protection are now a routine activity of states, but these often bely the difficulties inherent in determining a rationale for protection, which is essentially a philosophical problem (featuring debates over instrumental, intrinsic and the so-called 'eudaimonic' valuations, see, e.g., Chan et al., 2016), and those of establishing baselines against which ecological risk assessments can be made, which is both an applied science and an administrative problem (see, e.g., Jasanoff, 2013).

Social risk is applied both generally and specifically by different academic disciplines and individual studies, indicative of diverse interpretations of the key concepts of hazard, likelihood and vulnerability. In common usage, a 'social risk' basically refers to risks pertaining to people. Whereas across social science scholarship, governmental practices and in public policy, the term's meaning varies greatly, partly because of differences in how 'society' is understood in terms of composition, function, identity, institutions, relations with other groups and structure (and which, in turn, shapes interpretations of hazards and vulnerability). Across the disciplines of economics, geography, health studies, law, public administration and sociology are differing interpretations of social risk. Furthermore, some view social risks as those generated by society, i.e., as a *causal* source (such as by cities, industries, pollution or technology), while for others, social risks are those where society is exposed to risk from any source, i.e., as *consequences* (so that 'natural' disasters can generate social risks).

Social risk traditionally focuses on social policy and risks relating to social security (such as family disintegration, illiteracy, poverty and unemployment), socio-economic or social disturbances, such as conflicts (Armingeon & Bonoli, 2006), but increasingly includes impacts and hazards connected to global environmental change (UNISDR, 2015). Social risks often frame risk as stemming from the modernisation of society and, accordingly, have a clear focus on the effects on individuals, groups and organisations in society from different types of negative impacts (Beck, 1992; Crouch, 2015; Lash et al., 1996).

Even if there were scholastic agreement over the definition of social risk, it would remain controversial and contested as identifying or labelling a risk can have wide implications for communities, governments and industries. There is also an unhelpful tendency in some quarters to identify what are *social risk factors* as social risks; in other words, some views of social risks conflate causal factors and outcomes. Other problems bedevil pinning down a definition, such as that of subjectivity, as even technical risk assessments are shot through with judgements and assumptions based on values (see, e.g., Jasanoff, 2013). Drawing boundaries around risks can add to these problems, as uncertainties over characteristics, such as duration, extent and hazardousness, can cloud its identity. Consequently, social risk definitions are inherently conditional on circumstances and an outcome of political values, and the processes duly influenced by these values.

DIFFERENTIATING SOCIETAL RISK FROM SOCIAL RISK AND SYSTEMIC RISK

For this volume, societal risk needs to be differentiated from the more general term of social risk, as it is invested with a specific meaning and social risk will not suffice for this purpose. Moreover, applying societal risk to the problem of climate change evokes several deficiencies and limitations with the social risk concept. Social risk:

• Is so widely defined that it is best used only as a general description of risks in the social realm arising from all hazards,

- Has no agreed foundation as to the identity of society; indeed, it is usually applied without regard to scale and can refer to hazards to individuals, businesses and organisations, as well as to social units,
- Is typically uncertain in its understanding of the relationship between society and ecology, and
- Therefore, cannot differentiate between general harms to society and its constituent elements (including individuals), from risks to a society, including what might be termed existential risks to society.

Today, there is often talk of *systemic risks*, with the term 'systemic' referring to the perception that a risk, in principle, is always embedded in larger contexts of societal processes. Systemic risks have great potential for harm because these can be amplified or prolonged through system of interdependences (van Asselt & Renn, 2011). As such, systemic risks do not fit the traditional linear, mono-causal risk model (Renn et al., 2020), as they can be

> ...endogenous to, or embedded in, a system that is not itself considered to be a risk and is therefore not generally tracked or managed. Systems can contain latent, or cumulative, risk potential to impede overall system performance when some characteristics of the system change. (UNDRR, 2022, p. 4)

But a systemic risk is not the same as a societal risk. A focus on specific systems within society and Nature does not necessarily catch the complex interaction between different systems within society. Consequently, there is a need to develop a meaning and understanding of societal risk that transcends the limitations of ecological, social and systemic risk types and provides a basis for recognising the hazards of climate change in ways that have pragmatic and scholarly benefits.

Resolving, or at least reducing, these problems requires addressing some key issues, which are formulated as the following propositions:

- Societal risks focus on the threats to the social and socio-ecological systems on which society depends; risks to these systems are hazardous to society as an entity,
- Societal risks involve political values and ideologies, political actors and stakeholders, political institutions and political processes in risk perception, risk calibration and assessment, decision-making over

risk avoidance, mitigation and adaptations, and in 'do nothing responses',
- Societal risks include those to the environment, for reasons both instrumental and of recognising intrinsic values of ecological phenomena and processes,
- Societal risks from climate change can be addressed through GHG (greenhouse gases) mitigation, climate change adaptations, societal transformation and other contributory actions, and
- Societal risks from climate change cannot be resolved by individuals or corporations but are collective action problems requiring a collective response.

Therefore, societal risks can be given a specific meaning that distinguishes them from other types of risks.

APPLYING SOCIETAL RISK TO CLIMATE CHANGE RISKS

Climate change impacts can reach deeply into society, affecting not only socio-material networks and flows but also governance, lifestyles and politics in contemporary society. And as Chakrabarty (2009, p. 197) observes, climate change

> ...elicits a variety of responses in individuals, groups, and governments, ranging from denial, disconnect, and indifference to a spirit of engagement and activism of varying kinds and degrees.

Accordingly, climate change presents a collective action challenge for societal actors, permeated with uncertainty and without clearly identified borders or responsibilities (Moloney et al., 2018). Climate change is a truly complex societal challenge that, for good reasons, is labelled a 'diabolical policy problem' with a high probability of devastating outcomes for society (Steffen, 2012). Complexity here refers to climate changes' "...multiple driving forces, strong feedback loops, long time lags, and abrupt change behaviour" (Lenton et al., 2019, pp. 21–22). This also includes compound events combining interacting physical processes over multiple spheres and over spatial and temporal scales (Zscheischler et al., 2018).

Climate change also entails the possibility of what the IPCC labels 'large-scale singularities', namely abrupt and considerable changes in climate and climate-related systems (IPCC, 2019; Lenton et al., 2019). Potential threshold changes have been identified in the biosphere, cryosphere and in ocean/atmosphere systems. Of these relatively numerous potential events, a few have received particular attention: Irreversible melting of the Greenland and West Antarctica ice sheets, GHG releases (notably methane) from ground deposits due to permafrost thawing and the shutdown of the North Atlantic thermohaline ocean circulation. Climate change 'large-scale singularities' or 'tipping points' can relate to both ecological and social systems and entail irreversible damage to systems, potentially contributing to societal collapse (van Ginkel et al., 2020). This pertains not only to climate change impacts, but also to society's action to mitigate risks and undertake adaptations (Anguelovski et al., 2016; Sovacool, 2018).

Climate change as societal risk does not stand alone but is perceived as a broad risk that is increasingly integrated with related risks of environmental and social origins. Consequently, climate change risk responses have become incorporated into the policy and response measures in other risk fields, notably sustainable development, disaster risk reduction and resilience, such as in the three UN agreements of 2015; the Paris Agreement *on Climate Change* (UN FCCC, 2015), the *Sustainable Development Goals* (United Nations, 2015) and the *Sendai Framework for Disaster Risk Reduction 2015–2030* (UNISDR, 2015).

About This Book

Several conceptual boundaries are applied in this volume. *Firstly*, the conception of risks is that these are always a product of hazards. From the societal perspective, risks acquire their identity as a response to endangered values, so that risks are always additive, in that one risk does not cancel out another. Only if values change, may one risk replace another. Benefits (and co-benefits) arising from adaptive and other responses to hazards from climate change present as opportunities and are not direct risks; indeed, the volume recognises the potential social and environmental benefits of certain risk responses (including those achieved by societal transformation), *Secondly*, this volume is not concerned with the societal risks of GHG mitigation but with the risks associated with climate change and related impacts, *Thirdly*, the volume also does not specifically address the risks of adaptation (and maladaptive) responses, per se, although these are mentioned, *Fourthly*, this book deals with the contemporary climate change discourse, nominally dating from the latter 1970s, when the policy implications from anthropogenic climate change as a phenomenon become subject to international scientific and diplomatic debate. It is, however, informed by historical accounts and the analysis of climate-related events from pre-history, and *Fifthly*, although there is reference to financial (and economic) risks and to individual psychology and risk, neither of these specialist sub-fields are specifically addressed here.

The aims of this volume are as follows:

1. To identify the distinctive characteristics of climate change risks,
2. To review the understanding of the climate–society nexus,
3. To describe and review the major interpretations of risk,
4. To identify and define the societal risk of climate change, and
5. To consider contrasting futures for the climate–society nexus and the implications of climate as societal risk.

This volume is divided into four parts. Part I introduces the concept of climate change as societal risk and describes the distinctive risk characteristics presented by climate change (see Aims 1, 2 and 4). This chapter provides a primer on risk by addressing its key concepts and explaining the basic features of societal risk. Chapter 2 offers a temporal perspective of climate change risks, looking at the past, present and future to derive a proposition that climate change poses distinctive risks to societies.

Part II describes a range of theories used to understand risk, depicted as contrasting storylines, divided into scientific-technical and cultural/social risk theories (see Aim 3). Chapter 3 covers the scientific-technical theories, with the prominent theories of positivist theory, Normal Accident Theory and rational choice theory examined in turn, concluding with a critique of scientific-technical theories. Chapter 4 turns to the cultural/ social theories, examining cultural risk theory, risk society theory and the sustainability risk theory.

Part III discusses the concept of climate change as societal risk, and its implications, it is further developed, as informed by aspects of various storylines of risk described in the preceding part and the distinctive characteristics of climate change risks (see Aims 3, 4 and 5). Chapter 5 deals with societal risk in the context of specific features of climate change hazards. Chapter 6 concerns the discourse on climate and societal collapse and how this is used to formulate lessons relevant to the task of understanding societal risks in conditions of existential hazards. In the subsequent Chapter 7, the issue of climate change as societal risk in a political context is considered and its implications for societal transformation.

Part IV comprises the concluding chapter of the volume (Chapter 8), where the central arguments of the book are summarised by way of addressing the work's aims and drawing together the findings from the preceding chapters (see Aims 1, 2, 3, 4 and 5). Included also are some expositions and exegeses on the significance of the findings for policy formulation and associated research and scholarship.

REFERENCES

Adger, W. N., et al. (Eds.). (2009). *Adapting to climate change: Thresholds, values, governance*. Cambridge University Press.

Alexander, D. E. (2013). Resilience and disaster risk reduction: An etymological journey. *Natural Hazards and Earth System Sciences, 13*(11), 2707–2716. https://doi.org/10.5194/nhess-13-2707-2013

Anguelovski, I., et al. (2016). Equity impacts of urban land use planning for climate adaptation: Critical perspectives from the global north and south. *Journal of Planning Education and Research, 36*(3), 333–348. https://doi.org/10.1177/0739456x16645166

Armingeon, K., & Bonoli, G. (Eds.). (2006). *The politics of post-industrial welfare states: Adapting post-war social policies to new social risks*. Routledge.

Assmuth, T., et al. (2010). Integrated risk assessment and risk governance as socio-political phenomena: A synthetic view of the challenges. *Science of The Total Environment, 408*(18), 3943–3953. https://doi.org/10.1016/j.scitotenv.2009.11.034

Barnett, J., & Adger, W. N. (2007). Climate change, human security and violent conflict. *Political Geography, 26*(6), 639–655. https://doi.org/10.1016/j.polgeo.2007.03.003

Beck, U. (1992). *Risk society: Towards a new modernity*. Sage.

Campiglio, E., et al. (2018). Climate change challenges for central banks and financial regulators. *Nature Climate Change, 8*(6), 462–468. https://doi.org/10.1038/s41558-018-0175-0

Chakrabarty, D. (2009). The climate of history: Four theses. *Critical Inquiry, 35*(Winter), 197–222.

Chan, K. M. A., et al. (2016). Why protect nature? Rethinking values and the environment. *Proceedings of the National Academy of Sciences, 113*(6), 1462–1465. https://doi.org/10.1073/pnas.1525002113

Clayton, S., & Manning, C. (Eds.). (2018). *Psychology and climate change: Human perceptions, impacts, and responses*. Academic Press.

Crouch, C. (2015). *Governing social risk in post-crisis Europe*. Edward Elgar.

Crutzen, P., & Stoermer, E. (2000). The Anthropocene. *Global Change Newsletter, 41*, 17–18.

Douglas, M., & Wildavsky, A. (1983). *Risk and culture: An essay on the selection of technological and environmental dangers*. University of California Press.

Eckstein, D., et al. (2020). *Global climate risk index 2020: Who suffers most from extreme weather events? Weather-related loss events in 2018 and 1999 to 2018*. Germanwatch.

Fankhauser, S., & Schmidt-Traub, G. (2011). From adaptation to climate-resilient development: The costs of climate-proofing the millennium development goals in Africa. *Climate and Development, 3*(2), 94–113. https://doi.org/10.1080/17565529.2011.582267

GCA. (2019). *Adapt now: A global call for leadership on climate resilience*. Global Commission on Adapation.

Glover, L. (2006). *Postmodern climate change*. Routledge.

Glover, L., & Granberg, M. (2020). *The politics of adapting to climate change*. Palgrave Macmillan.

Hartzell-Nichols, L. (2017). *A climate of risk: Precautionary principles, catastrophes, and climate change*. Routledge.

IFRC. (2020). *World disaster report 2020: Come heat or high water*. International Federation of Red Cross and Red Crescent Societies. https://media.ifrc.org/ifrc/world-disaster-report-2020

IPCC. (2019). *Global warming of 1.5°c*. Intergovernmental Panel on Climate Change (IPCC).

IPCC. (2022). *Climate change 2022: Impacts, adaptation and vulnerability*. Intergovernmental Panel on Climate Change (IPCC).

IPCC. (2023). *Synthesis report of the sixth assessment report (AR6)*. Intergovernmental Panel on Climate Change (IPCC).

Jasanoff, S. (2013). *Science and public reason* (Paperback ed.). Earthscan.

Kasperson, J. X., & Kasperson, R. E. (Eds.). (2005). *The social contours of risk* (Vol. I). Earthscan.

Lash, S., et al. (Eds.). (1996). *Risk, environment & modernity*. Sage.

Lawrence, J., et al. (2020). Cascading climate change impacts and implications. *Climate Risk Management, 29*, 100234. https://doi.org/10.1016/j.crm.2020.100234

Leichenko, R., & O'Brien, K. (2019). *Climate and society: Transforming the future*. Polity Press.

Lenton, T. M., et al. (2019). Climate tipping points—Too risky to bet against. *Nature, 575*(7784), 592–595. https://doi.org/10.1038/d41586-019-035 95-0

Levin, S. A., et al. (2022). Governance in the face of extreme events: Lessons from evolutionary processes for structuring interventions, and the need to go beyond. *Ecosystems, 25*(3), 697–711. https://doi.org/10.1007/s10021-021-00680-2

Lopéz-Carresi, A., et al. (Eds.). (2014). *Disaster management: International lesson in risk reduction, response and recovery.* Routledge.

Lupton, D. (Ed.). (1999). *Risk and sociocultural theory* (2nd ed.). Cambridge University Press.

Malloy, J. T., & Ashcraft, C. M. (2020). A framework for implementing socially just climate adaptation. *Climatic Change, 160*(1), 1–14. https://doi.org/10. 1007/s10584-020-02705-6

Meerow, S., et al. (2019). Social equity in urban resilience planning. *Local Environment, 24*(9), 93–808. https://doi.org/10.1080/13549839.2019. 1645103

Moloney, S., et al. (2018). Climate change responses from the global to the local scale: An overview. In S. Moloney, et al. (Eds.), *Local action on climate change: Opportunities and constraints* (pp. 1–16). Routledge.

Nicholls, R. J., et al. (1995). Impacts and responses to sea-level rise: Qualitative and quantitative assessments. *Journal of Coastal Research*, 26–43.

Pelling, M. (2011). *Adaptation to climate change: From resilience to transformation.* Routledge.

Pidgeon, N., et al. (Eds.). (2003). *The social amplification of risk.* Cambridge University Press.

Reisinger, A., et al. (2020). *The concept of risk in the IPCC sixth assessment report: A summary of cross-working group discussions.* Intergovernmental Panel on Climate Change (IPCC).

Renn, O., et al. (2020). Systemic risks from different perspectives. *Risk Analysis, n/a*(n/a), 1–19. https://doi.org/10.1111/risa.13657

Schlosberg, D., et al. (2017). Adaptation policy and community discourse: Risk, vulnerability, and just transformation. *Environmental Politics, 26*(3), 413–437. https://doi.org/10.1080/09644016.2017.1287628

Sloat, L. L., et al. (2020). Climate adaptation by crop migration. *Nature Communications, 11*(1), 1243. https://doi.org/10.1038/s41467-020-150 76-4

Slovic, P. (2000). *The perception of risk.* Routledge.

Sovacool, B. K. (2018). Bamboo beating bandits: Conflict, inequality, and vulnerability in the political ecology of climate change adaptation in Bangladesh. *World Development, 102*, 183–194. https://doi.org/10.1016/j.worlddev. 2017.10.014

Steffen, W. (2012). A truly complex and diabolical policy problem. In J. S. Dryzek, et al. (Eds.), *The Oxford handbook of climate change and society* (pp. 21–37).

Steffen, W., et al. (2018). Trajectories of the earth system in the anthropocene. *Proceedings of the National Academy of Sciences, 115*(33), 8252–8259. https://doi.org/10.1073/pnas.1810141115

Taleb, N. N. (2008). *The black swan: The impacts of the highly improbable.* Penguin Books.

Tulloch, J. (2008). Culture and risk. In J. O. Zinn (Ed.), *Social theories of risk and uncertainty: An introduction* (pp. 138–167). Blackwell.

UNDRR. (2022). *Global assesment report on disaster risk reduction 2022.* United Nations Office for Disaster Risk Reduction (UNDRR).

UN FCCC. (2015). *Paris agreement.* United Nations Framework Convention on Climate Change (UN FCCC).

UNISDR. (2015). *Sendai framework for disaster risk reduction 2015–2030.* United Nations Office for Disaster Risk Reduction (UNISDR).

United Nations. (2015). *Transforming our world: The 2030 agenda for sustainable development.* United Nations.

Vale, L. J. (2014). The politics of resilient cities: Whose resilience and whose city? *Building Research & Information, 42*(2), 191–201. https://doi.org/10.1080/09613218.2014.850602

van Asselt, M. B. A., & Renn, O. (2011). Risk governance. *Journal of Risk Research, 14*(4), 431–449. https://doi.org/10.1080/13669877.2011.553730

van Ginkel, K. C. H., et al. (2020). Climate change induced socio-economic tipping points: Review and stakeholder consultation for policy relevant research. *Environmental Research Letters, 15*(2), 023001. https://doi.org/10.1088/1748-9326/ab6395

van Valkengoed, A., & Steg, L. (2019). *The psychology of climate change adaptation.* Cambridge University Press. https://doi.org/10.1017/9781108595438

Winkelmann, R., et al. (2022). Social tipping processes towards climate action: A conceptual framework. *Ecological Economics, 192,* 107242. https://doi.org/10.1016/j.ecolecon.2021.107242

Zscheischler, J., et al. (2018). Future climate risk from compound events. *Nature Climate Change, 8*(6), 469–477. https://doi.org/10.1038/s41558-018-0156-3

CHAPTER 2

A Climate of Risks

Abstract This chapter delves into the identity and character of climate and climate change risks. These risks are socially constructed and, as such, respond to dynamic changes in society. By examining climate and climate change risks from three contrasting temporal perspectives, the shifts in these risks can be brought to light. Historical, contemporary and future manifestations of climate change and climate risks are examined in turn. Climate change as a 'threat multiplier' that exacerbates pre-existing hazards and risks through environmental degradation and depletion of the earth's resources is explained. The chapter finishes with a discussion of the distinctive character of climate risks.

Keywords Civilisation · Climate change · Climate change indicators · Collapse · History · Temporal perspectives · Threat multiplier

Climate is fundamental to life, but this benevolence is paired with its dangers and those of climatic change (Lenton et al., 2020). Prior to the age of science, weather events harmful to society and the character of the climate straddled cultural notions of the natural and the supernatural. In our modern era, climate and climate change can be analysed for the risks they present to society and to ecological values, the irony of which is the human imprint on climatic change as an agent of climate

© The Author(s), under exclusive license to Springer Nature   25
Switzerland AG 2023
M. Granberg and L. Glover, *Climate Change as Societal Risk*,
https://doi.org/10.1007/978-3-031-43961-2_2

change. A feature of life in the Anthropocene is that the global climate system and social life are mutually influential, with society having acquired the role of influencing the climate system. Natural history and human history have become one (Chakrabarty, 2009). Through declarations of climate change as a societal risk, we are describing the potential dangers of the climate. Of these dangers, extreme weather events and their consequences may be more newsworthy and impose extreme mortality and morbidity losses on human populations, however other (slower-onset) climatic changes are equal or greater threats if essential socio-ecological systems could be disrupted or lost.

The debate over climate change features apocalyptic visions of the consequences of global warming and these are part of the debate over what constitutes dangerous climate change (Schnellnhuber et al., 2006). A great deal of this discussion concentrates on the rate and extent of future warming and its associated effects on the global climate and related systems. Apocalyptic accounts often carry, either explicitly or implicitly, an association with civilisational collapse, and studies of civilisational collapse due to climatic factors are used as exemplars of the social dangers of climate change (Brozović, 2023).

Globalisation has extended the social and economic interconnectiveness of all manner of social and commercial enterprises. Consequently, severe disruptions in one location may not be confined by cultural, national or geographic boundaries but might produce catastrophic events in other locations and in unanticipated ways. While specifying the quantified extent of future warming as being dangerous is useful in political and publicity campaigns, it is of limited utility—if not highly problematic—for assessing societal risk. There are several reasons for this limited usefulness:

- Societal impacts from climate change are highly differentiated by economic, geographical, political, social and other factors,
- Climate change presents as a very distinctive set of changes occurring over different time scales, so that extreme events are mixed with slower-onset changes (that can be *slow-moving disasters*), and
- Climate change impacts in the social and ecological realms can be extremely difficult to forecast, due to uncertainties in such factors as normal variability and social preferences, subjectivity in assessing vulnerability and systems with threshold responses.

Additionally, scholarly attention in the climate change discourse has turned to the problem of *cascading disasters* (Lawrence et al., 2020). Pescaroli and Alexander (2016) argue that there is a special class of cascading disaster that results from interdependencies in complex social and ecological systems. There is a strong societal element here, as cascading disasters "...tend to highlight unresolved vulnerabilities in human society" (2016, p. 65). Cascading uncertainties are another source of problems in efforts to quantify dangerous levels of global warming, as there is a long causal and circumstantial chain between the effects of rising global GHG emissions and possible local impacts (see, e.g., Schneider & Kuntz-Duriseti, 2002).

An area of increasing concern is the climatic risks intersecting with other trends that amplify societal risk and expand the realm of potential harms and losses. Increasing globalisation is one source of this threat, as: (1) There is greater interconnectivity between nation states and their economies, such that local losses or disruptions can be magnified into far-reaching and even global events, (2) That there is diminishing cultural diversity worldwide which reduces potential resilience to external shocks, and (3) Local independence and self-determination in lesser-developed nations and for weaker economic agents is reduced through increased economic connectivity to the global economy, thereby diminishing opportunities for developing local resilience. Although globalisation has boosted the wealth of many nations and alleviated much mass poverty, it has also increased global consumption of renewable and non-renewable resources through the spread of mass consumption societies. This is, of course, a part of the 'Great Acceleration' in populations, economic activity and resource use, culminating in the Anthropocene (see, e.g., Steffen et al., 2015).

Climate change risks are primarily those negative consequences created by the impacts of climate change; however, there is also an additional category of risk arising from maladaptation (Barnett & O'Neill, 2010). Many of these negative effects have been identified and described in climate change scholarship covering adverse outcomes in the cultural, economic and social realms. Of these negative effects, many occur from complex interactions between many social and environmental systems, with exemplars including damage or loss to (see, e.g., IPCC, 2023):

- Agriculture, forestry, primary industry and natural resource harvesting,
- Assets, infrastructure and property,

- Cities and settlements,
- Cultural materials,
- Ecosystem services,
- Human health, and
- Water system supply.

Such interdependencies pose a challenge to a narrow science-based climate analysis of risks and potential hazards.

Looking at how societal risks change over time is instructive in understanding the dynamic aspects of these risks. Three temporal perspectives on these dangers can be seen ranging across a variety of disciplinary backgrounds: (1) Historical and pre-historical, (2) Contemporary accounts, and (3) Forecasts and speculations on the future. The degree and extent to which the temporal perspectives are empirically based varies considerably.

LOOKING BACK

That civilisations organically grew, blossomed and then withered was a staple view in archaeology and history—as was the notion that these ends could be sudden and catastrophic (Chakrabarty, 2009). Societal collapse was typically deemed to be a combination of economic, political and social causal factors. Detrimental environmental change, including climate change, is also recognised as a contributor in some collapses. Until the contemporary era of the climate change discourse however, climate did not feature strongly in most accounts of collapse. Since then, climatic change has become a prominent factor and cited as an antecedent for the future (Brozović, 2023; Diamond, 2005; Tainter, 1988).

Numerous anthropological and historical studies examine the harmful effects of extreme weather events and climatic change in pre-history and history. Societal collapse is presented as the most dramatic of the dangers of climate change (as discussed below in Chapter 6), especially when deemed to be the determinant of that collapse. Although societal collapse has been typically explained using a combination of factors, in recent times, many studies describe climate change as its major cause, spanning a wide range of civilisations. Zheng et al. (2014), for example, describe the role of climate in the Ming dynasty's collapse through the social consequences of aridification and desertification, and declines in grain production and severe droughts. Li et al. (2016) find that climate change may be responsible for the (rise and) fall of the oasis communities along the Silk Road., while Buckley et al. (2010) conclude that prolonged drought brought on the demise of Angkor, the capital of Cambodia's Khmer Empire. Collapsed societies of particular scholarly interest also include Mexico's Maya, the ancient Egyptians of the Nile River and the Babylonians on the Tigris and Euphrates Rivers. An edited collection by Weiss (2017) considers the role of megadroughts in causing the collapse of civilisations ranging from foraging in Western Asia 12,000 bp, to the late Bronze Age collapse, to the Angkor in the fourteenth and sixteenth centuries.

Climate history covers a wider array of climatic influences than those contributing to societal collapse, and, in total, these histories inform contemporary understandings of the societal risks of climate change (Chakrabarty, 2009). Explorations of the effects of the Little Ice Age, for example, range widely and identify its influence in agricultural production and economic activity and extensively in the social and cultural realms of political and military developments (Mann, 2002). Of particular interest here, is the historical climate influence on society, especially how knowledge and culture contribute to earlier societies' recognising climatic risks and to shaping their responses.

Looking Around

In 1988, NASA climatologist James Hansen, famously stated: "It's time to stop waffling … and say that the greenhouse effect is here and is affecting our climate" (quoted in, Bowen, 2008, p. 1). Climate change is not something that awaits us in a distant future and giving us ample time to make decisions and adapt society to its impacts. Furthermore, the global climate will continue to change in harmful ways (Hartzell-Nichols, 2011). In August 2021, the IPCC released a report that concludes that climate change is an ongoing, human-induced hazard that manifests itself in increased frequencies and magnitudes of cyclones, droughts, extreme temperatures, fires, heavy precipitation and compound extreme events (2021). It is well underway, therefore, and provides insights into the implications of the impacts of more extreme future climatic change (IPCC, 2023). The harmfulness of future effects "…will depend both on the extent to which we are able to mitigate further greenhouse-gas (GHG) emissions, and on how well we can adapt to our changing climate" (Hartzell-Nichols, 2014, p. 150).

For some commentators, however, it is already too late to arrest the prospect of dangerous climate change. Philosopher Dale Jamieson (2014, p. 12) opined:

> As a result of climate change, there will be massive extinctions of plants and animals, rising seas will engulf major cities and entire nations, and "natural" disasters including droughts, heat waves, and storms will raise havoc with virtually all aspects of life. The landscape in which people live will be remade by climate change.

A feature of the impacts of climate change is the sheer extent of the effects, spanning the globe and influencing an immense range of social activities. Categories of impacts from the IPCC impacts report (2022) indicate the breadth of the ecosystems, ecosystem services and social systems and societal values effected: terrestrial and freshwater ecosystems; oceans and coastal ecosystems; water; food, fibre and other ecosystem products; cities, settlements and infrastructure; health, well-being and communities; poverty, livelihoods and sustainable development. Particular attention is paid to unique and threatened keystone species, the frequency and scale of extreme weather events and the effects of climate change on disadvantaged communities. Human health is a major concern; some experts consider climate change to be the biggest global human health challenge of the twenty-first century (Costello et al., 2009; Quinn et al., 2023). As grave as these global risks are, in a sense, the extent of societal risk cannot be properly appreciated at aggregated scales, given the extent of variation and complexity of climate change impacts at the local levels and between different sectors.

Under the accelerated program of climate monitoring, modelling and analysis in the contemporary era, a considerable body of knowledge of the changes underway in climatic and related systems has been established (IFRC, 2020; IPCC, 2023; WMO, 2021). Every passing year in contemporary times brings forth new insights into the ways in which climate is changing and of the growing and increasing implications for the natural and social worlds and the connections between them (see Box 2.1). A precise reckoning, however, of what we have lost so far due to climate change, as a global society and in environmental values, is unavailable.

Box 2.1. Selected climate change indicators

GHG concentrations: In 2019, atmospheric carbon dioxide concentrations reached 410 parts per million (148% higher than pre-industrial levels), methane 1877 parts per billion (260% higher) and nitrous oxide 332 parts per billion (123% higher).

Temperature: Global mean temperature in 2020 was 1.2 °C above the 1850–1900 baseline; 2020 was one of the three warmest years on record.

Global mean sea levels: These have risen with 3.3 mm annually since 1993 and continue to do so, with regional variations; ocean heat content at 0–200 m depth continued to rise in 2019 and did also at greater depths; marine heatwaves deemed 'strong' occurred at least once across much of the ocean in 2020; ocean acidification trends continued in 2020; deoxygenation trends are uncertain but oxygen levels have declined in the open oceans by 0.5–3% since 1950.

In the cryosphere: Annual minimum sea ice extent in the Arctic was the second lowest on record in 2020; those 40 glaciers with long-term observations experienced ice-loss in 2018–2019 close to the record levels of 2017–2018.

Source WMO (2021)

In our time, catastrophic weather events are regularly in the news around the world and record-breaking weather is similarly commonplace; the exceptional has become the routine (IFRC, 2020). From climatic natural hazards alone, the costs are high; according to the Climate Risk Index (Eckstein et al., 2021), some 475,000 lives were lost between 2000 and 2019 as a direct consequence of over 11,000 extreme weather events.

Such extreme weather events (notably cyclones/hurricanes, droughts, floods, heatwaves, forest fires and bushfires) demonstrate the potential for extensive and intensive losses to net human welfare and these losses are typically highly differentiated according to social vulnerability. Occurrences of such events in 2020 and 2021 offer insights into the future risks and harms possible under a changing climate (see, e.g., Table 2.1).

Table 2.1   A selection of high-impact events, 2020 and 2021

Event	Locale
Widespread Flooding—Significant loss of life	Kenya, Sudan
Widespread Flooding—Loss of life, Significant population displacement	Benin, Burkina Faso, Cameroon, Chad, Côte d'Ivoire, Ethiopia, Kenya, Niger, Nigeria, Senegal, Somalia, South Sudan, Sudan, Togo, Uganda
Monsoon season—Significant loss of life	Afghanistan, Bangladesh, India, Myanmar, Nepal, Pakistan
Severe flooding	Brazil, Democratic Republic of the Congo, Indonesia, Rwanda, Vietnam, Yemen
Severe Drought	Argentina, Brazil, Paraguay, Uruguay
Significant/Severe Forest and Bushfires	Argentina, Brazil, Paraguay
Severe Forest and Bushfires	Australia, USA
Long-term Drought Continuance	Southern Africa
Tropical Cyclones	Indian Ocean: Major losses/disruptions in Bangladesh, India North Atlantic: Cuba, Dominican Republic, Haiti, Honduras Nicaragua, USA Pacific Ocean: Major losses in Fiji, Philippines, Vanuatu

Source WMO (2021)

Risks connected or related to climate change are of great concern given the extent and scale of potential social harm and loss, especially for global warming at the higher range scenarios. Low-lying coastal areas and small island developing states are vulnerable to coastal floods, cyclones/hurricanes and sea-level rise, especially the world's densely populated deltas (including the Ganges, Indus, Mekong, Nile and Yangtze). Extreme weather events can cause long-lasting economic losses in settlements and rural areas and damage to critical infrastructure (including that for communications, education, health and transport), in addition to direct losses from morbidity and mortality. Disruption to agricultural and fisheries production threatens food supply and food security, with great implications for those communities, households and locations already experiencing stress under existing conditions. A key aspect of such agricultural disruptions with implications for settlements are threats to water supply, both to harvesting quantities and to reliability. Damage and loss to ecosystems will invariably cause loss of environmental values, notably accelerated biodiversity loss, and to ecosystem goods and services, an impact with widespread implications for human welfare.

As IPCC (2022) describes, in addition to climate change's ecological and physical impacts, interacting social factors influence social vulnerability to its impacts, with cultural, economic, institutional and social marginalisation increasing vulnerability, arising from factors such as discrimination based on age, class, culture, ethnicity, gender and other identity and social attributes. Therefore, the impacts of climate change on society and its individuals are mediated through social structures.

LOOKING FORWARD

There has been no shortage of apocalyptic climate visions of the future in recent times; what separates many of these from similar earlier accounts is their scientific basis (Skrimshire, 2014; Smith, 2022; Swyngedouw, 2010). Whereas earlier apocalyptic warnings were often intended to motivate religious, political and social reactions, scientific forecasts are now describing such outcomes as essentially being assured—the question is not 'If?', but 'When, and how bad?' In 2007, Mark Lynas published *Six Degrees* that brings together scientific findings of the effects of increments of global warming from 1 °C to 6 °C and describes the escalation in social and environmental losses, while noting that warming so far has already reached 1 °C. Seemingly alarmist at the time, similar projections to Lynas's made more recently have become familiar and now attract little controversy. Documenting the research advances in impacts forecasting by the IPCC's report series (see, 2022, 2023) describes climate change impacts across a wide array of social and ecological values, where obviously the greater the global warming, the greater the loss of these values.

The IPCC states that the limit of 1.5 °C warming will probably be reached by 2040 (see, 2021). Current international policies will be inadequate to prevent temperature rises of 2.8 °C by the end of the century (UNEP, 2022). It is too late to do anything about the effects of the GHG emissions to date, as these have entered the climate system and the effects will be in place for millennia, regardless of the future GHG emissions trajectory. Anthropogenic climate change is producing adverse effects on society and ecology and will continue to do so over time; the risks of climate change have been realised and these hazards can only continue and worsen (IPCC, 2022).

As disturbing as the extent of these potential impacts are, two factors escalate these risks. *Firstly*, these impacts could be far worse than forecast due to tipping points in the climate and related natural systems (Steffen et al., 2018). As Lenton (2011) describes, although components of earth systems could produce *large-scale singularities*, these were viewed as high risk/low probability outcomes, but scientific re-assessments of global warming suggest these probabilities may not be low (i.e., the possibility of a high risk/high probability). *Secondly*, climate change acts in concert with other global environmental issues and societal risk, therefore, is the sum effects of these influences. Steffen et al. (2004), describe the character of global change in the earth system as interactions of natural systems and human activities. Such confluences produce multiple climate hazards and the phenomena of compounding and cascading risks. Climate risks are also connected to socio-economic tipping points, materialising through societal disruptions and large economic damages that can lead to non-linear and abrupt changes in social systems "...to a fundamentally different state, following relatively marginal climate change" (van Ginkel et al., 2022, p. 1). Although climate change fits into much of the general discourse of risk assessment and analysis, it has several distinguishing characteristics that need to be taken into account when considering societal risk.

DISTINCTIVE CHARACTER OF CLIMATE RISKS

A number of factors contribute to the distinctive character of climate change risks.

Perception. As contemporary history shows, in modern societies in a global economy, climate risk does not impress itself immediately or prominently into consciousness (Harris, 2013; Urry, 2011). Its character defies ready perception, where more fundamental climatic change is usually hidden within daily weather, where modern societies depend on distant industrial agricultural output traded in global markets and where exogenous energy sources and systems and technological innovations are employed to smooth out the discontinuities and disruptions in local socio-ecological systems. Awareness of the causes of anthropogenic climate change is obscured as the association between the release of pollutants and their effect on the global climate can only be assuredly known through advanced measurement, scientific research and advanced computation.

Although contemporary responses to climate change are but the latest chapter of this relationship, extant climate change is of decidedly different character to that of the past due to its anthropogenic causes (Rosenzweig & Neofotis, 2013). Only with the passage of time and with developments in computation, monitoring, science and technology is society now able to comprehend the risks associated with anthropogenic climate change. In this sense, risk is the connective tissue between the trilogy of climate change discourse by linking the rationales for investigating the physical science of climate change, climate change impacts and climate change adaptations.

IPCC (2022) identifies those risk-increasing factors as a crossing of system thresholds associated with even more accelerated changes, extreme events, greater magnitude and rate of change, unanticipated changes (i.e., surprises) and the extent of future global warming at the higher end of estimates (i.e., greater climate sensitivity to increased GHG concentrations). Added to these are the aforementioned 'non-linear, complex and discontinuous' changes in response to climate change that may result in "...substantially greater sensitivity to further stimulus or dramatic change, explosive growth, or collapse" (IPCC, 2001, p. 93). Therefore, it is important to identify risks arising and from interactions between responses (Andrews et al., 2023), especially those relating to compound climate events (Simpson et al., 2023). Response risks can include responses to climate change that fail to "...achieve their intended outcomes, as well as responses creating additional adverse outcomes as they exacerbate hazards, vulnerability, and exposure to climate change risk" (Andrews et al., 2023, p. 1). Ambiguities over climate change impacts in local and regional climate change scenarios, and the range of possible impacts combined, amplify this uncertainty. At the base of these factors is the driver of future climatic change, namely ongoing and future (anthropogenic) GHG emissions, the rate and scale of which influences the extent, persistence, type, scale and timing of future climate change impacts (see, IPCC, 2023).

Analysis and management. Climate change risks challenge the relevant analysis and management tasks, spanning problem framing, risk assessment, risk management and risk communication (IFRC, 2020; IPCC, 2023; UNDRR, 2019). To begin, the timeframes involved can be lengthy; decisions made today can have effects persisting for very long periods. In the popular media, the prominent risks of climate change are the potential losses (and actual losses) from climate change impacts, especially when combined with interconnected physical system changes (such as sea-level rise) and when combined with other, sometimes interconnected, impacts of other human activities (such as biodiversity loss, monoculture agriculture, pollution and urbanisation). There are, however, serious climate risks that have received little public attention and may be of great importance, such as the climate system tipping points (Lenton et al., 2019). It follows that the probability ranges for climate risks vary greatly between climate phenomena, locations and socio-ecological systems. Cascading and compounding risks present particular difficulties for prediction, assessment and adaptation and planning responses. Climate risks are also dynamic, changing in response to anticipatory measures of adaption, awareness-raising and prevention.

Risk response. How society responds to these climate risks also generates its own risks, although many of these have been overlooked or under-appreciated (Andrews et al., 2023; Simpson et al., 2023) (see Table 2.2). Hence, identifying, evaluating and debating the risks of responding to climate change are a sizable portion of the climate change discourse at local, city, state, national and international levels. Proposals that GHG emissions could be mitigated by replacing fossil fuel electricity generation with a revitalised nuclear power sector evoked public controversy over the risks of nuclear energy, for example (see, Helm, 2011; Hultman, 2011). Suggestions that large-scale sequestration through monoculture plantations (and bioenergy with carbon capture and storage) could provide a means to lower atmospheric concentrations of carbon drew criticism over the potential economic, environmental and social costs (cf., Haikola et al., 2019). Similarly, large-scale geoengineering works to manipulate the climate system, such as aerosol injections to the upper atmosphere, cloud brightening, ocean fertilisation to promote algal growth and surface albedo changes, carry potentially great risks (cf., Jean Buck, 2012; Schneider, 1996). Adaptation measures, of any scale, taken in anticipation or response to climate change impacts also entail risk-taking and possible maladaptation. To these risks, must be added the risks of deliberately deciding not to adapt (Nicholls et al., 1995). Policymakers and other relevant decision-makers need also to consider not only the risks of the outcomes of these responses (as known costs and unintended consequences) but also those of the failure of the action or measure to fulfil its intended purpose (Andrews et al., 2023).

Subjectivity. A wide range of scholarly disciplines are engaged in the task of describing and predicting climate change impacts in the physical,

Table 2.2 Primary and direct/indirect risks of climate change

Risk source	Overview
Impacts of Climate Change	Losses and damages from actual or forecast climate change, including lost/ diminished opportunities
GHG Emissions Mitigation Measures	Hazards created or increased by emissions mitigation actions
Carbon Cycle Measures & Climate (and Related Systems) Geoengineering	Hazards created or increased by measures to lower atmospheric GHGs concentrations and/or the conduct of large-scale geoengineering projects of climate (and related systems) management
Adaptation Responses to Climate Change	Hazards or increased or created by adaptation responses and through maladaptation
Inaction in the Face of Known Risks	Risks of not taking action to known risks (i.e., the risks of 'do nothing' responses)

social and eco-social realms, thereby addressing *what* is at risk, *where* it will occur and the timeframes over *when* this will occur and prevail, based on available knowledge (cf., Pidgeon & Fischhoff, 2013). Largely, these investigations are technical and scientific in character and method and exhibit the virtues associated with scientific knowledge. Turning to the questions of the social aspects of the risks of climate change—the *who, why* and *how* inquiries—more challenging aspects of scholarship are evoked because these inquiries clearly have a political identity (cf., Glover & Granberg, 2020). See, for example, Hultman et al. (2010, p. 287) who query the UN FCCC goal of preventing dangerous anthropogenic inter-ference with the climate system (in Art. 2) by asking 'Dangerous to whom?' They point out the perpetual dangers of climate and weather to vulnerable peoples and raise the following:

The implicit, but questionable, statement is that current levels of human adaptation to climate variability and extreme weather are considered roughly acceptable, whereas those projected for a warming climate are not.

As the authors conclude, this determination cannot be made objectively but must involve a value judgement and, therefore, "…even though it can be usefully informed by science, it cannot be determined by science" (2010, p. 287).

Indirect risks: Climate change can also produce potential side effects with cascading impacts on social systems (Pescaroli, 2018). As the frequency and intensity of extreme climatic events increases—such as heatwaves, catastrophic cyclones, forest fires, landslides, intense rainfall and floods—there are direct and indirect consequences affecting socio-economic and socio-ecological conditions (Khatun et al., 2022). Oftentimes, the effects of disasters spread far beyond the immediate disaster site, such as a drought in one nation can cause food shortages in another, a disaster causing economy-wide disruptions, or businesses and industries unrelated to a disaster site being harmed because capital, labour or material flows are interrupted or re-directed. Repercussions of climate change on land and resources on which human populations depend can have effects on social migratory patterns (IPCC, 2021). As many harmful climate and related hazards are materialising more often and at greater scales, "…migration has erupted as a distinct discourse, with copious developmental, humanitarian, research, and policy fields aiming to grasp its complexity" (Silchenko & Murray, 2023, p. 2); (migration is explicitly examined in Chapter 5).

Another indirect risk, partly related to migration, concerns food security, arising from disruptions to food availability, distribution, production and sustainability (HLPE, 2020). Determinants of food security, as for many other types of security, are

> …as temporally as they are spatially complex: past processes … shape present insecurities, and ongoing processes such as climate change and trade liberalisation shape future insecurities … larger scale processes that shape people's entitlements … may themselves be vulnerable to climate change. (Barnett & Adger, 2007, p. 642)

Climate change impacts can lead to a breakdown of food systems, including those for crops, fisheries or livestock, or as disruptions in food distribution due to drought, flooding and precipitation variability and other weather extremes (Mirzabaev et al., 2023). This indirect climate risk is a particular threat to vulnerable populations, due to factors such as their low-income status, inadequate access to nutritious food and social discrimination. People's exposure to heatwaves can also affect their livelihoods and incomes (de Lima et al., 2021). Climate change's impact on resources central to people's and societies' sustenance also involves the indirect risk of violent conflict, itself a cause of human vulnerability to climate change (Barnett & Adger, 2007). Although the causal chain in these indirect effects is often recognised and understood, a feature of disasters is their capacity to reveal the importance of hitherto unrealised systemic links through failure and for which no planning or preparations have been made. Arguably, this is a feature of nearly every major natural (and technological) disaster.

New risks. Climate change-related impacts alter the character and magnitude of known climate risks and also generate new risks for social, economic, technological and ecological systems (Dow et al., 2013). Notions that society can be independent of the natural world must contend with the complex, dense and intertwined relationship between the social realm and the influences of climate and its related systems in the society–climate nexus. Change in the society–climate nexus can originate either in society or climate. Further to this relationship is that research into climate and society continues to uncover new aspects of these relationships, serving to underscore the reliance of society on different ecosystems (Wainwright & Mann, 2020). As adaptation to climate change impacts increases and evolves, the connection to risk management becomes more evident and there is also a "…greater recognition that the rate and magnitude of climate variability and change may exceed the limits to adaptation of socio-ecological systems" (Dow et al., 2013, p. 384).

As this chapter illustrates, climate change is a 'threat multiplier' exacerbating pre-existing hazards and risks through environmental degradation and depletion of the earth's resources (CNA, 2007). The UN science advisory body, the IPCC, provides an overview of the interlinked challenges of radically reducing the dependency on fossil fuel and preparing for future outbreaks of extreme weather (2021). Threats caused by climate change create risks also evoke the logics of securitisation and protectionism (Young, 2003). Yet, in terms of specific societal risks, the connection between concern over the degree of changes in the climate and the social consequences has often been unclear. On the other hand, there are a great number of studies of specific activities, locations and phenomena analysing the potential social and environmental impacts of climate change. This is something of a classic top-down problem of analysis, whereby it is difficult to reconcile the local with the general (Krauß & Bremer, 2020). Accordingly, there is a need to understand the social (and linked ecological) risks at scales between the global/national and the local/household and consider the risks to societies (i.e., the unit of the societal).

REFERENCES

Andrews, T. M., et al. (2023). Risk from responses to a changing climate. *Climate Risk Management, 39,* 100487. https://doi.org/10.1016/j.crm.2023.100487

Barnett, J., & Adger, W. N. (2007). Climate change, human security and violent conflict. *Political Geography, 26*(6), 639–655. https://doi.org/10.1016/j.pol geo.2007.03.003

Barnett, J., & O'Neill, S. J. (2010). Maladaptation. *Global Environmental Change, 20*(2), 211–213.

Bowen, M. (2008). *Censoring science: Inside the political attack on dr. James Hansen and the truth of global warming* (Paperback ed.). Plume.

Brozović, D. (2023). Societal collapse: A literature review. *Futures, 145,* 103075. https://doi.org/10.1016/j.futures.2022.103075

Buckley, B. M., et al. (2010). Climate as a contributing factor in the demise of Angkor, Cambodia. *Proceedings of the National Academy of Sciences, 107*(15), 6748–6752. https://doi.org/10.1073/pnas.0910827107

Chakrabarty, D. (2009). The climate of history: Four theses. *Critical Inquiry, 35*(Winter 2009), 197–222.

CNA. (2007). *National security and the threat of climate change.* The CNA Corporation.

Costello, A., et al. (2009). Managing the health effects of climate change. *The Lancet, 373*(9676), 1693–1733.

de Lima, C. Z., et al. (2021). Heat stress on agricultural workers exacerbates crop impacts of climate change. *Environmental Research Letters, 16*(4), 044020. https://doi.org/10.1088/1748-9326/abeb9f

Diamond, J. (2005). *Collapse: How societies choose to fail or survive.* Viking.

Dow, K., et al. (2013). Limits to adaptation to climate change: A risk approach. *Current Opinion in Environmental Sustainability, 5*(3), 384–391. https://doi.org/10.1016/j.cosust.2013.07.005

Eckstein, D., Künzel, V., & Schäfer, L. (2021). *The global climate risk index 2021.* Germanwatch.

Glover, L., & Granberg, M. (2020). *The politics of adapting to climate change.* Palgrave Macmillan.

Haikola, S., et al. (2019). From polarization to reluctant acceptance–bioenergy with carbon capture and storage (beccs) and the post-normalization of the climate debate. *Journal of Integrative Environmental Sciences, 16*(1), 45–69. https://doi.org/10.1080/1943815X.2019.1579740

Harris, P. G. (2013). *What's wrong with climate politics and how to fix it.* Polity.

Hartzell-Nichols, L. (2011). Responsibility for meeting the costs of adaptation. *WIREs Climate Change*, 2(5), 687–700. https://doi.org/10.1002/wcc.132

Hartzell-Nichols, L. (2014). Adaptation as precaution. *Environmental Values*, 23(2), 149–164. https://doi.org/10.3197/096327114X13894344179121

Helm, D. (2011). Nuclear power, climate change, and energy policy. In D. Helm & C. Hepburn (Eds.), *The economics and politics of climate change* (Paperback ed., pp. 247–262). Oxford University Press.

HLPE. (2020). *Food security and nutrition: Building a global narrative towards 2030*. High Level Panel of Experts on Food Security and Nutrition of the Committee on World Food Security (HLPE).

Hultman, N. E. (2011). The political economy of nuclear energy. *WIREs Climate Change*, 2(3), 397–411. https://doi.org/10.1002/wcc.113

Hultman, N. E., et al. (2010). Climate risk. *Annual Review of Environment and Resources*, 35(1), 283–303. https://doi.org/10.1146/annurev.environ.051308.084029

IFRC. (2020). *World disaster report 2020: Come heat or high water*. International Federation of Red Cross and Red Crescent Societies. https://media.ifrc.org/ifrc/world-disaster-report-2020

IPCC. (2001). *Climate change 2001: Impacts, adaptations and vulnerability*. Cambridge University Press.

IPCC. (2021). *Climate change 2021: The physical science basis*. Intergovernmental Panel on Climate Change (IPCC).

IPCC. (2022). *Climate change 2022: Impacts, adaptation and vulnerability*. Intergovernmental Panel on Climate Change (IPCC).

IPCC. (2023). *Synthesis report of the sixth assessment report (ar6)*. Intergovernmental Panel on Climate Change (IPCC).

Jamieson, D. (2014). *Reason in a dark time: Why the struggle against climate change failed—And what it means for our future*. Oxford University Press.

Jean Buck, H. (2012). Geoengineering: Re-making climate for profit or humanitarian intervention? *Development and Change*, 43(1), 253–270. https://doi.org/10.1111/j.1467-7660.2011.01744.x

Khatun, F., et al. (2022). Environmental non-migration as adaptation in hazard-prone areas: Evidence from coastal Bangladesh. *Global Environmental Change*, 77, 102610. https://doi.org/10.1016/j.gloenvcha.2022.102610

Krauß, W., & Bremer, S. (2020). The role of place-based narratives of change in climate risk governance. *Climate Risk Management*, 28, 100221. https://doi.org/10.1016/j.crm.2020.100221

Lawrence, J., et al. (2020). Cascading climate change impacts and implications. *Climate Risk Management*, 29, 100234. https://doi.org/10.1016/j.crm.2020.100234

Lenton, T., M, et al. (2020). Life on earth is hard to spot. *Anthropocene Review*, 7(3), 248–272. https://doi.org/10.1177/2053019620918939

Lenton, T. M. (2011). Early warning of climate tipping points. *Nature Climate Change, 1*, 201–209.

Lenton, T. M., et al. (2019). Climate tipping points—Too risky to bet against. *Nature, 575*(7784), 592–595. https://doi.org/10.1038/d41586-019-035 95-0

Li, Z., et al. (2016). Drought promoted the disappearance of civilizations along the ancient silk road. *Environmental Earth Sciences, 75*(14), 1116. https://doi.org/10.1007/s12665-016-5925-6

Lynas, M. (2007). *Six degrees: Our future on a hotter planet.* 4th Estate.

Mann, M. E. (2002). The earth system: Physical and chemical dimensions of global environmental change. In M. MacCracken & J. S. Perry (Eds.), *Encyclopedia of global environmental change.* John Wiley & Sons.

Mirzabaev, A., et al. (2023). Severe climate change risks to food security and nutrition. *Climate Risk Management, 39*, 100473. https://doi.org/10.1016/j.crm.2022.100473

Nicholls, R. J., et al. (1995). Impacts and responses to sea-level rise: Qualitative and quantitative assessments. *Journal of Coastal Research*, 26–43.

Pescaroli, G. (2018). Perceptions of cascading risk and interconnected failures in emergency planning: Implications for operational resilience and policy making. *International Journal of Disaster Risk Reduction, 30*, 269–280. https://doi.org/10.1016/j.ijdrr.2018.01.019

Pescaroli, G., & Alexander, D. (2016). Critical infrastructure, panarchies and the vulnerability path of cascading disasters. *Natural Hazards, 82*, 175–192. https://doi.org/10.1007/s11069-016-2186-3

Pidgeon, N., & Fischhoff, B. (2013). The role of social and decision sciences in communicating uncertain climate risks. In J. Arvai & L. Rivers III (Eds.), *Effective risk communication.* Routledge.

Quinn, T., et al. (2023). Health and wellbeing implications of adaptation to flood risk. *Ambio.* https://doi.org/10.1007/s13280-023-01834-3

Rosenzweig, C., & Neofotis, P. (2013). Detection and attribution of anthropogenic climate change impacts. *WIREs Climate Change, 4*(2), 121–150. https://doi.org/10.1002/wcc.209

Schneider, S. H. (1996). Geoengineering: Could—or should—we do it? *Climatic Change, 33*(3), 291–302. https://doi.org/10.1007/BF00142577

Schneider, S. H., & Kuntz-Duriseti, K., et al. (2002). Uncertainty and climate change policy. In S. H. Schneider (Ed.), *Climate change policy: A survey* (pp. 53–87). Island Press.

Schnellnhuber, H. J., et al. (Eds.). (2006). *Avoiding dangerous climate change.* Cambridge University Press.

Silchenko, D., & Murray, U. (2023). Migration and climate change—The role of social protection. *Climate Risk Management, 39*, 100472. https://doi.org/10.1016/j.crm.2022.100472

Simpson, N., et al. (2023). Adaptation to compound climate risks: A systematic global stocktake. *SSRN*. https://doi.org/10.2139/ssrn.4205750

Skrimshire, S. (2014). Climate change and apocalyptic faith. *WIREs Climate Change, 5*(2), 233–246. https://doi.org/10.1002/wcc.264

Smith, C. (2022). Climate change and culture: Apocalypse and catharsis. *Ethics & the Environment, 27*(2), 1–27.

Steffen, W., et al. (2015). The trajectory of the anthropocene: The great acceleration. *The Anthropocene Review, 2*(1), 81–98. https://doi.org/10.1177/2053019614564785

Steffen, W., et al. (2018). Trajectories of the earth system in the anthropocene. *Proceedings of the National Academy of Sciences, 115*(33), 8252–8259. https://doi.org/10.1073/pnas.1810141115

Steffen, W., et al. (2004). *Global change and the earth system: A planet under pressure*. Springer.

Swyngedouw, E. (2010). Apocalypse forever? *Theory, Culture & Society, 27*(2–3), 213–232. https://doi.org/10.1177/0263276409358728

Tainter, J. A. (1988). *The collapse of complex societies*. Cambridge University Press.

UNDRR. (2019). *Global assessment report on disaster risk reduction 2019*. United Nations Office for Disaster Risk Reduction (UNDRR).

UNEP. (2022). *Emissions gap report 2022: The closing window*. UNEP.

Urry, J. (2011). *Climate change & society*. Polity.

van Ginkel, K. C. H., et al. (2022). A stepwise approach for identifying climate change induced socio-economic tipping points. *Climate Risk Management, 37*, 100445. https://doi.org/10.1016/j.crm.2022.100445

Wainwright, J., & Mann, G. (2020). *Climate Leviathan: A political theory of our planetary future* (Paperback ed.). Verso.

Weiss, H. (Ed.). (2017). *Megadrought and collapse: From early agriculture to Angkor*. Oxford University Press.

WMO. (2021). *State of the global climate 2020*. World Meteorological Organization (WMO).

Young, Iris M. (2003). The logic of masculinist protection: Reflections on the current security state. *Signs: Journal of Women in Culture and Society, 29*(1), 1–25. https://doi.org/10.1086/375708

Zheng, J., et al. (2014). How climate change impacted the collapse of the Ming dynasty. *Climatic Change, 127*(2), 169–182. https://doi.org/10.1007/s10584-014-1244-7

PART II

Contrasting Risk Theories for Interpreting the Climate-Society Nexus

CHAPTER 3

Scientific-Technical Theories of Risk

Abstract Opening with a brief history of risk theory, the chapter then examines scientific-technical risk theories. Three key theories are covered: Normal accident theory, Positivist risk theory and Rational choice risk theory. These reviews encompass how risk is defined and understood as a scientific phenomenon, the view that catastrophic failures in complex technological systems are inevitable and how individual rationality is perceived as the basis for societal decision-making on risk. An overview of the critique of scientific-technical risk theory is given and the chapter closes with an account of the relevance of scientific-technical risk theory to societal risk.

Keywords History of risk · Normal Accident Theory · Positivist risk theory · Rational choice risk theory · Scientific-technical risk theory

Although risk is not often considered as such, it may be regarded as being in part a metaphysical problem, in that human life (indeed, all life as we understand it) is subject to forces external to our being and beyond our control. Pre-industrial societies did not experience risks as these are understood in the contemporary manner of modern societies, where technological control over Nature was limited and where

> ...the uncertainties of everyday life were expressed and managed through a range of religious and magical beliefs in concepts such as fate, providence and luck ... During the medieval period, such beliefs were part of a cosmology in which every earthly event and the fate of every individual was depicted as a symbolic representation of the will of God. (Reith, 2004, p. 386)

As per the quote from Shakespeare's *Hamlet*: "There's a divinity that shapes our ends, rough hew them how we will". One does not need to believe in divine providence to recognise the ubiquity of fate in life's course. Risks from natural hazards were deemed to be an 'act of God' and this perception of risk excluded the idea of human agency, responsibility and meaningful intervention. Critically, and as the quote eloquently captures, counterposed against fatalistic determinism is the impulse, indeed necessity, to seek to control our destinies in an uncertain world. Modernity's emergence through the Enlightenment, after which industrialisation and capitalism established a new world order and, as a result, societies developed "...a system of strategies and beliefs in the attempt to deal with, contain and prevent danger" (Lupton, 1999, p. 3). Today, decisions of individuals and organisations are considered the root causes of disasters and, at the same time, decisions in society about the future are dominated by ideas about risks (Luhmann, 1993). Consequently, it has been stated that, if we are under the sway of modernity, we live in a *risk society* (see, e.g., Beck, 1992).

Over the course of this evolution in human perception, the concept of risk expands from having its locus in Nature to now including human agency and the functioning of society (Lupton, 1999, 2013). Risk from this perspective, therefore, bridges from Nature to the individual and organisational levels of society, highlighting its societal dimensions. In modern, western society, and those societies sufficiently under its influence, the concept of risk

Table 3.1   Scientific-technical risk theories

Risk theory	Key characteristic
Positivist Risk Theory	Risk as defined and understood as a scientific phenomenon
Normal Accident Theory	Catastrophic failures in complex technological systems are inevitable
Rational Choice Risk Theory	Individual rationality as the basis for societal decision-making on risk

…is widely used to explain deviations from the norm, misfortune and frightening events. This concept assumes human responsibility and that 'something can be done' to prevent misfortunes. (Lupton, 2013, p. 3)

Risk analysis has its roots in this historical change in perception, but as a specific field of scientific-technical activity, it began when complicated engineering systems—such as factories for chemical production, nuclear power plants and other complex and multifaceted technological systems with potential for disastrous mishaps—became more commonplace (McDaniels, 2021). Through the logic and experience of modernity, a technocratic and apolitical perspective of risk was established, guiding policy and measures for risk reduction and management. Of many competing conceptualisations of risk, three essential scientific-technical risk theories are reviewed below (see Table 3.1).

POSITIVIST RISK THEORY

In modern societies, conceptualisations of risk focus on scientific facts as the basis for decision making. This emphasises the role of technical and rational processes, and bureaucratic systems, in enabling hazard prevention by identifying threats beforehand and by subsequently managing these threats (Bradbury, 2009). Such processes rely heavily on a positivist tradition of science based on logical empiricism, a realist-objectivist ontology and the rational actor paradigm (Rosa, 1998; Shrader-Frechette, 1991). In one sense, all social risk responses stem from risk perception and, in modern and industrial societies, that perception is largely informed by scientific-technical monitoring, assessment and evaluation (Lupton, 2013).

In the public policy discourse of experts, senior public sector managers and administrators and public agencies, corporations and many business operators, and others dealing with public risk management in official capacities, scientific-technical risk approaches dominate (Douglas, 1992; Luhmann, 1993; Pidgeon et al., 2003). Such practices are conducted by professionals with scientific and technical expertise; examples of scientific-technical approaches include science research and risk analyses into defence, ecological risks, engineering, infrastructure, occupational health and safety, and security (for an interesting overview of the history of scientific assessment in environmental policy, see, Oppenheimer et al., 2019). An obvious explanation for these circumstances is the sheer scale of social life dealing with individual and societal risk. Furthermore, because of the centrality of risk in the roles and activities of representative governments in modern states, there is a considerable body of awareness-raising campaigns, laws and regulations, policy, plans, public information, research, scholarship and standards, all based on scientific-technical risk discourse.

Risk, in this modern industrial world, "...could be described, quantified and predicted and managed or avoided" (Lupton, 2013, p. 7). From this orientation, risk is viewed as the probability of negative outcomes that could be realised in relation to a specific hazard, but usually depicted as the consequences of the pursuit or continuance of a benefit, gain or need (Bradbury, 2009). Prominent in this discourse, therefore, are the themes of the dichotomy of risk/reward, the probability of outcomes and applying this knowledge in rational modes of viewing decision-making at a collective scale (such as deployed in response strategies, such as avoidance, risk minimisation, risk transfer and adaptation).

For many of the professionals and public officials involved in risk assessment and risk management, it is assumed that risk can essentially be conceptualised as a scientific- technical concept (Head, 2014). It would be an unexceptional claim in these groups to equate societal risk with a metric that measures the relationship between the frequency of an event and the extent of harm suffered by a population (often using metrics such as morbidity/mortality and economic damages and losses). In many instances, there are established metrics and procedures for their use to be followed in calculating standards and in conducting risk assessments required by law or regulation. 'Natural disaster' responses involving protecting public interests at large are vested in public agencies. Actions, plans and policies from these agencies are responses built from the scientific-technical knowledge of these 'natural disaster' risks and conform to the same logic. Risk identification, analysis and evaluation precede risk treatments and subsequently, monitoring and review.

NORMAL ACCIDENT RISK THEORY

Charles Perrow analyses a set of well-known disasters in *Normal Accidents* (1984) and formulates what has become known as Normal Accident Theory (NAT); later additions added more recent disasters (1999). His assertion that accidents are to be expected, such that accidents are normal occurrences, challenges the conventional presumptions that highly dangerous technologies can be designed and operated safely. Perrow argues that in highly complex technological systems (characterised as having 'interactive complexity' and being 'tightly coupled'), accidents are inevitable—and that accidents in these systems will become more frequent. Complexity applies to technologies, and sometimes to the organisations, involved. Disasters covered in the earlier work include those in nuclear power plants, refineries and shipping. Importantly, under NAT, technological systems with these characteristics will be subject to disasters that can be neither predicted nor prevented.

Normal Accidents is a highly quoted work in the field of disaster studies and remains highly influential in policy, practice and research (Silvast & Kelman, 2013). NAT generated controversy, as its basic premise runs counter to many of the tenets of organisational theory, professional risk management and the engineering and design principles of large technological systems. By implication, NAT undermines the social contract over the development and use of these systems.

Although Perrow (1984, 1999) offers ways to reduce the risks of these systems, it is the very existence of the devices, practices and procedures that adds to system complexity and the 'coupling' within systems, that makes accidents more likely. 'Coupling' has a particular meaning that applies to the necessity of following exact sequences of steps within specified time periods in response to existing circumstances. But increasing system complexity obscures decision-makers' awareness of prevailing conditions and risks. Introducing redundancy in systems and adding safety devices can be counter-productive, not only by adding to complexity, and adding elements that can themselves fail. These additional safety measures can also influence decision-making by encouraging greater risk-taking (thereby evoking the *risk compensation* theory). Expressing

NAT in terms of risk, efforts at risk reduction through these means produce the unintended outcome of *increasing* the hazardousness of these systems.

Perrow's work (1984, 1999) can be seen now as either prompting or supporting the change in the ways that disasters in complex systems are analysed and approached, shifting the focus from failures by individuals, technical glitches or sheer happenstance, to considering system failures that cause operators to make catastrophic errors. Risk management has moved from attempting to prevent negligence and human failings, to that of dealing with systemic failures and system deficiencies. There is a further and deeper implication of normal accidents, as Perrow concluded of the Fukushima nuclear accident:

> Some complex systems with catastrophic potential are just too dangerous to exist, not because we do not want to make them safe, but because, as so much experience has shown, we simply cannot. (2011, p. 52)

NAT has barely registered in climate discourse, and this may be an oversight. *Firstly*, although other risk perspectives address technological issues in climate change, notably the technology risks of nuclear power and geoengineering, few have elected to draw (explicitly or knowingly) on the insights of a concept that directly applies to the central concerns over technology safety. *Secondly*, most risk perspectives have a neutral position on technologies or are uninterested in their place and role, yet in contemporary modern societies, technology is interwoven throughout the activities of hazard identification and risk assessment, management and communication. Arguably, decision-makers would be better informed by possessing knowledge on the risks of available technology choices. *Thirdly*, although NAT is not a critique of science and technology per se, it questions how corporations and public institutions generate, interpret and apply such knowledge to their own ends, thereby providing the foundation of a counter-narrative to 'official' views on technology risks.

Certainly, NAT concerns can be found in implicit form in the climate change discourse. In considering cascading effects and cascading risks, Pescaroli and Alexander (2016, p. 188) state

> ...the vulnerability of critical infrastructure cannot be related only to the built environment but must also be connected to the different levels of responsibility and human interaction: pre- and post-emergency

management, urban and infrastructural planning, and political and social visions.

Michael Hulme (2014, p. 112), for example, rejects geoengineering solutions to global climate change as a dangerous experiment, stating: "It is not possible to know what the consequences of such engineering would be". Here, NAT takes the form that geoengineering of the global climate constitutes a reckless experiment and one of excessive risk.

RATIONAL CHOICE RISK THEORY

Rational choice risk theory applies individual utilitarianism to decision theory and posits, in the simplest form, that individuals make decisions based on their assessments of the need to best serve their ideal interests and objectives (Ostrom, 1990). Maximising one's self-interest is based on anticipations and expectations of future outcomes which, in the case of risk, is the optimum avoidance of future loss from options known (or expected) to be available. It is taken that this individual rationality is also that of groups and collective enterprises. There is a considerable intellectual lineage to this view, from J.S. Mill and the utilitarian philosophers, with Adam Smith and the rationality of the 'invisible hand' in economic market systems, to contemporary neoliberalism in economics and politics being prominent elements (cf., Cannavò & Lane, 2014).

Promoting self-interest would carry little moral suasion as a social ethos without the accompanying claim that the collective effect of individual self-satisfaction produces optimal net social outcomes (Scott, 2000). In other words, the net effect of individual 'rationality' is the maximisation of societal rationality and, therefore, the production of net social benefit. Rational choice theory takes economic principles and applies these in analysing an array of social phenomena involving various resources, some of which are (relatively) intangible social values.

Rational choice finds explicit political expression in neoliberalism, which has arguably been the dominant ideology of most contemporary governments around the world and many international organisations with interests in economic policy and development since the 1980s (Cohen, 2012; Fieldman, 2011; Lockie, 2013). Neoliberalism promotes a political program evoked by its utilitarian ideology, expressed largely through economic policy instruments in the public sector, and as actively promoted by conservative political groups, private enterprises and associated NGOs with an interest in its fulfilment. With its influence spread so widely and deeply, neoliberalism impinges upon the risk discourse and practices in disaster risk and sustainable development, and subsequently the climate change discourse.

Those advocating a minimal role for government in climate change responses usually promote rational choice responses and, as Fankhauser et al. (1999, p. 74) put it, "Perhaps the main role for government will be to provide the right legal, regulatory and socio-economic environment to support autonomous adaptation". Mendelsohn (2000) offers that governments can undertake measures to benefit affected stakeholders that might, if acting independently, disregard the risks to others.

Exemplifying this approach, Filatova (2014) considers a range of market-based instruments for dealing with the climate change impacts from floods as a way of avoiding the costs and limitations of conventional land use planning. Government efforts at flood risk management and government investments in public infrastructure, it is suggested, encourage occupation and further private investment in flood risk locations. Using market-based instruments—insurance, marketable permits, subsidies and taxes—offers an alternative to government 'command-and-control' planning by removing market distortions and incentives to occupy flood-prone areas and encourages autonomous adaptations by private parties. Central to this argument is the reallocation of risk by placing greater responsibilities onto individuals and firms.

CRITICISING SCIENTIFIC-TECHNICAL RISK THEORIES

Limits to, and Values of, Positivism

Although climate change risks are now understood from many perspectives and engage many scholarly disciplines, the essential precursor to this varied enterprise was that of disaster research. For the first fifty years or so of disaster research (until around the turn of the last century), the prevailing view was that knowledge for disaster reduction should be built on rational, detailed scientific investigation and research. Support for this positivist approach began to wane, with some critical observers calling it 'a delirium of rationality' (Castel, 1991, p. 289). As Ismail-Zadeh et al. (2017) argue, positivism fosters increases in deeper single-discipline knowledge—but limits the appreciation of other management options.

Up to this period, risk did not feature strongly in social sciences scholarship, with sociology being the obvious exception. Public controversies, however, over industrial pollution, nuclear energy and other phenomena, prompted dissatisfaction with probabilistic risk calculations framed within technical discourses (see, e.g., Lupton, 1999, Chapter 2), leading to the rise of psychometric analysis of risk, drawing on cognitive and psychological roots of individual risk perception (Slovic, 1992; Slovic et al., 1986). Yet, these reforms kept risk recognition and assessment firmly within the positivist tradition.

Concerns continued over the scientific framing of risk, based partly on political and ideological objections, notably over elite and authoritarian decision-making, cultural biases of decision-makers and the phenomenon of *technocracy* (Klinke & Renn, 2019; Lupton, 1999; Löfstedt & Boholm, 2009; Villagrán de León, 2012; Zinn, 2008). Fundamentally, expert-oriented scientific-technical risk models misalign with perceptions among individuals and groups and can differ considerably from expert judgements, due to cultural factors, information availability and media coverage (Kahneman, 2012; Slovic, 2000).

Appreciating the behavioural, economic, political and social dimensions of risks and risk responses was deemed the corrective course. Subsequently, risk analysis broadens out into new realms to incorporate different social perspectives on risk, which entails drawing on cultural, political and socio-economic factors. Applying the repair of multidisciplinary contributions, however, is insufficient to satisfy the more fundamental critiques based on questioning scientific rationality as the model of risk knowledge, risk discourse and associated decision-making processes (Hirsch Hadorn et al., 2008).

When confronting politics, occurring when risk management issues spill into the public realm, the authority of science and experts can be considerably decreased when compared to risk management as confined to the activities of governments and their agents (Jasanoff, 2013; Kasperson et al., 2005; Siegrist et al., 2009). Public controversy can be expected to accompany most major risk issues, especially those of a technological character. Issues that bring scientific-technical appraisals into public conflict in public risk issues vary greatly and include:

• Accusations of the politicisation of science by vested interests,
• Cultural insensitivity,
• Disputations over claims of understated public risks,

- Inequitable or unfair reasoning in risk estimation, or outcomes, in risk management proposals, and
- Neglect of specific environmental and social values in risk assessment.

Politics, regardless of the views or preferences of rational decision-makers, are invariably and inexorably inherent in decisions regarding risk involving collective decision-making. Anthropological, political and sociological studies identify the influence of politics and culture in creating risk, risk exposure and risk responses and the effectiveness of any responses (Bulkeley, 2021; Crate, 2011; Nursey-Bray & Palmer, 2018; Paulson et al., 2003; Stripple & Bulkeley, 2013). A political critique is also the basis of many objections to scientific-technical approaches to risk assessment as the primary tool for guiding the responses to risk, especially by public agencies and those governing risks in the social sphere (such as regulatory oversight of the civil and private sectors) (Jasanoff, 1986, 2005).

Fundamentally, science's claims to objectivity were refuted, thereby bringing into play the notion that scientific risk assessment entails judgements and further, when subsequently applied in policy decisions, the outcomes could legitimately claim to have been made objectively. Cultural, political and power dimensions of knowledge creation and sense-making are highlighted in this critique (Forsyth, 2009; Karvonen & Brand, 2009). Subsequently, there were demands for greater accountability by decision-makers, inclusion of knowledge inputs from a wider array of stakeholders, that attention be paid to the limits of available knowledge and that there be more transparency in risk assessment processes.

Science Versus Anti-science

Questioning the use of science for resolving social issues is not the same as rejecting science, per se. In an era where science is spurned, at least in selected applications, by those espousing *anti-science*, it may be useful to briefly distinguish between the broader social anti-science movement and the scholarly critique of scientific rationality. Especially so, since the climate change discourse has featured a highly influential, politically motivated and well-funded contribution from climate change sceptics and deniers (see, Dunlap & McCright, 2013; Oreskes & Conway, 2010). This is often part of a conservative political agenda that has a

"...staunch commitment to free markets and disdain of governmental regulations" (Dunlap & McCright, 2013, p. 144). Anti-science movements are also closely connected to populism, with a strong transformative agenda for challenging the status quo and introducing 'new' political conflict dimensions (Mudde, 2004, 2017). 'Popular' anti-science sentiments tend to have their social foundations in one or more of the following (Oppenheimer et al., 2019; Oreskes, 2019):

- Scientific 'illiteracy' and belief in *pseudoscience*
- Where scientific findings conflict with organised/personal spiritual views or articles of faith, and
- Where social groups realise scientific findings conflict with their political beliefs/programs, particularly where science is held to be supporting cosmopolitan, modernist or progressive social causes.

Scientific scepticism is also cultivated by professionally managed campaigns aiming to protect vested private interests, typically by large corporations or their consortiums, such as the case with climate change. Expressions of anti-science arise in constructed controversies over science, where science from established and legitimate institutions is contested by 'outsiders' employing discredited scientific findings or fabricated knowledge in public campaigns over singular issues (Dunlap & McCright, 2013). Anti-science advocates in such issues typically have no reservations over their personal use of science and technology, or its use in their campaigns. Those with deep concerns over modern technologies *in toto* are rare and may be followers of deep ecology philosophies. Complaints or disputes over scientific findings, critiques of science (or scientists) and objections to the politicisation of science, are not, in isolation, anti-science positions and in many instances are efforts to protect the legitimacy of the scientific enterprise and its outputs.

Conceptual and Related Weaknesses in Normal Accident Theory

NAT attracts scholarly criticism, much of it stemming from ambiguities in its central ideas. These objections include that it:

- Does not constitute a theory, but refers to a loose category of events (Hopkins, 1999) and that it is 'un-falsifiable' (Silvast & Kelman, 2013),
- Relies on central concepts of 'coupling' and 'complexity' that are vague and inconsistent, and
- Is fatalistic and is (negatively) technologically deterministic.

Countering NAT is the high-reliability theory that argues that complex systems can be, and indeed are, managed safely. In this theory, safety is achieved by using shared decision-making in organisations that prioritise safety and in systems that have ongoing learning, system redundancy and established communication channels (see, e.g., Grabowski & Roberts, 1997; LaPorte & Consolini, 1991; Weick et al., 1999). In this debate, Marais et al. (2004) observe the importance of the source of risk, pointing out that the risks in technological systems are not only technological, but also social and organisational; these are among those safety researchers promoting a systems approach to safety.

Political Values in Rational Choice Theory

Proponents of rational choice freely mix political and economic arguments to the extent that rational choice should be viewed as primarily a political program utilising economic concepts as the foundations of a free market ideology (see, e.g., Anderson & Leal, 2001). Criticisms of rational choice decision-making stem, accordingly, from ideological and related concerns (see, Green & Shapiro, 1994), namely that rational choice:

- Both describes an observable practice and depicts an ideal,
- Embodies unrealistic key assumptions, including that information on all choices is freely and equally available, is presented without bias and that decisions are made without coercion or deception,
- Ideologically, is an amalgam of economic, political, psychological and sociological elements and is neither neutral (i.e., objective) nor 'scientific', and
- Models collective behaviour and choice are without reference to scale; social behaviour is taken to be the sum of individual behaviours (i.e., 'methodological individualism').

Other critics have taken issue with the 'rationality' aspects of the theory. In short, these include rejecting or questioning (Douglas, 1992; Lupton, 1999, 2013):

- That individual and social risk responses follow a predictable path because they are universally guided by the dictates of self-interest and 'rational' behaviour,
- That the pivotal rational 'economic man' (sic) concept faithfully represents human decision-making and is ideologically neutral or without variability in judgement, and
- That other factors do not interfere with ideal rationality, such as imperfect information and deception and coercion.

Relevance of Scientific-Technical Risk Theory to Societal Risks of Climate Change

Science is the bedrock of our knowledge of global climate change and climate change science is one of the great achievements in the progress of human knowledge. Important as the manufactured political campaigns opposing climate change science have been, the focus of the inquiries in this volume are with mainstream climate science. Scientific knowledge is essential in identifying the impacts of climate change, but it is incomplete and potentially misleading in understanding the risks of these impacts in the social and related spheres.

Scientific-technical approaches to risk cannot replace or substitute for social representations and constructions of risk. Accordingly, the climate sciences' emphasis on quantitative and probabilistic precision is problematic, not least when the aim is to communicate risk (Bloomfield & Manktelow, 2021; Shepherd et al., 2018). It is, therefore, important to acknowledge that narratives/stories are central for interpreting and evaluating the way we live our lives and this puts a focus on non-quantifiable local climate narratives about a changing climate and how this is understood (Baztan et al., 2020; Krauß & Bremer, 2020; Marschütz et al., 2020).

Climate change impacts have great local variations and so do local narratives and understandings that contribute to the problem framing of climate change and associated risks (Carr & Nalau, 2023). Past events also need to be included, as they "...involve detailed stories which can be unpacked" (Shepherd et al., 2018, p. 557). Hence, local climate narratives can "...give meaning to abstract scientific information" (Krauß & Bremer, 2020, p. 2) and need to be taken into account, as the positivist paradigm inhibits "...the social dynamics of environmental change" (Marín-Puig et al., 2022, p. 2). This is a strong argument for challenging the dominant modes of technocratic risk governance and its role in political action, and this opens the way for social and cultural risk theories, as reviewed in the next chapter.

References

Anderson, T. L., & Leal, D. R. (2001). *Free market environmentalism*. Palgrave.
Baztan, J., et al. (2020). Facing climate injustices: Community trust-building for climate services through arts and sciences narrative co-production. *Climate Risk Management, 30*, 100253. https://doi.org/10.1016/j.crm.2020.100253
Beck, U. (1992). *Risk society: Towards a new modernity*. Sage.
Bloomfield, E. F., & Manktelow, C. (2021). Climate communication and storytelling. *Climatic Change, 167*(3), 34. https://doi.org/10.1007/s10584-021-03199-6
Bradbury, J. A. (2009). The policy implications of differing concepts of risk. In R. E. Löfstedt & Å. Boholm (Eds.), *The Earthscan reader on risk* (pp. 27–42). Earthscan.
Bulkeley, H. (2021). Climate changed urban futures: Environmental politics in the anthropocene city. *Environmental Politics, 30*(1–2), 266–284. https://doi.org/10.1080/09644016.2021.1880713
Cannavò, P. F., & Lane, J. H. (Eds.). (2014). *Engaging nature: Environmentalism and the political theory canon*. The MIT Press.
Carr, E. R., & Nalau, J. (2023). Adaptation rationales and benefits: A foundation for understanding adaptation impact. *Climate Risk Management, 39*, 100479. https://doi.org/10.1016/j.crm.2023.100479
Castel, R., et al. (1991). From dangerousness to risk. In G. Burchell (Ed.), *The Foucault effect: Studies in governmentality* (pp. 281–298). The University of Chicago Press.
Cohen, A. (2012). Rescaling environmental governance: Watersheds as boundary objects at the intersection of science, neoliberalism, and participation. *Environment and Planning A, 44*, 2207–2224.
Crate, S. A. (2011). Climate and culture: Anthropology in the era of contemporary climate change. *Annual Review of Anthropology, 40*(1), 175–194. https://doi.org/10.1146/annurev.anthro.012809.104925
Douglas, M. (1992). *Risk and blame: Essays in cultural theory* Routledge.

Dunlap, R. E., & McCright, A. M., et al. (2013). Organized climate change denial. In J. Dryzek (Ed.), *The Oxford handbook of climate change and society* (pp. 3–17). Oxford University Press.

Fankhauser, S., et al. (1999). Weathering climate change: Some simple rules to guide adaptation decisions. *Ecological Economics, 30*(1), 67–78. https://doi.org/10.1016/S0921-8009(98)00117-7

Fieldman, G. (2011). Neoliberalism, the production of vulnerability and the hobbled state: Systemic barriers to climate adaptation. *Climate and Development, 3*(2), 159–174. https://doi.org/10.1080/17565529.2011.582278

Filatova, T. (2014). Market-based instruments for flood risk management: A review of theory, practice and perspectives for climate adaptation policy. *Environmental Science & Policy, 37*, 227–242. https://doi.org/10.1016/j.envsci.2013.09.005

Forsyth, T. (2009). Democratizing environmental expertise about forests and climate. In G. Kütting & R. Lipschutz (Eds.), *Environmental governance: Power and knowledge in a local-global world* (pp. 170–185). Routledge.

Grabowski, M., & Roberts, K. (1997). Risk mitigation in large-scale systems: Lessons from high reliability organizations. *California Management Review, 39*(4), 152–161. https://doi.org/10.2307/41165914

Green, D. P., & Shapiro, I. (1994). *Pathologies of rational choice theory: A critique of applications in political science.* Yale University Press.

Head, B. W. (2014). Evidence, uncertainty, and wicked problems in climate change decision making in Australia. *Environment and Planning C: Government and Policy, 32*(4), 663–679. https://doi.org/10.1068/c1240

Hirsch Hadorn, G., et al. (Eds.). (2008). *Handbook of transdisciplinary research.* Springer.

Hopkins, A. (1999). The limits of normal accident theory. *Safety Science, 32*(2), 93–102. https://doi.org/10.1016/S0925-7535(99)00015-6

Hulme, M. (2014). *Can science fix climate change?* Polity.

Ismail-Zadeh, A. T., et al. (2017). Forging a paradigm shift in disaster science. *Natural Hazards, 86*(2), 969–988. https://doi.org/10.1007/s11069-016-2726-x

Jasanoff, S. (1986). *Risk management and political culture.* Russell Sage Foundation.

Jasanoff, S. (2005). Judgment under siege: The three-body problem of expert legitimacy. In S. Maasen & P. Weingart (Eds.), *Democratization of expertise?* (pp. 209–224). Springer.

Jasanoff, S. (2013). *Science and public reason* (Paperback ed.). Earthscan.

Kahneman, D. (2012). *Thinking fast and slow.* Penguin Books.

Karvonen, A., & Brand, R. (2009). Technical expertise, sustainabilty, and the politics of special knowledge. In G. Kütting & R. Lipschutz (Eds.), *Environmental governance: Power an knowledge in a local-global world* (pp. 38–59). Routledge.

Kasperson, R. E., et al. (2005). Social distrust as a factor in siting hazardous facilities and communicating risk In J. X. Kasperson & R. E. Kasperson (Eds.), *The social contours of risk. Volume I* (pp. 29–50). Earthscan.

Klinke, A., & Renn, O. (2019). The coming of age of risk governance. *Risk Analysis, 41*(3), 544–557. https://doi.org/10.1111/risa.13383

Krauß, W., & Bremer, S. (2020). The role of place-based narratives of change in climate risk governance. *Climate Risk Management, 28*, 100221. https://doi.org/10.1016/j.crm.2020.100221

LaPorte, T. R., & Consolini, P. M. (1991). Working in practice but not in theory: Theoretical challenges of "high-reliability organizations". *Journal of Public Administration Research and Theory: J-PART, 1*(1), 19–48. http://www.jstor.org/stable/1181764

Lockie, S., et al. (2013). Neoliberalism by design: Changing modalities of market-based environmental governance. In S. Lockie (Ed.), *Routledge international handbook of social and environmental change* (pp. 70–80). Routledge.

Luhmann, N. (1993). *Risk: A sociological theory*. Aldine de Gruyter.

Lupton, D. (Ed.). (1999). *Risk and sociocultural theory* (2 ed.). Cambridge University Press.

Lupton, D. (2013). *Risk* (2 ed.). Routledge.

Löfstedt, R. E., & Boholm, Å. (2009). The study of risk in the 21st century. In R. E. Löfstedt & Å. Boholm (Eds.), *The Earthscan reader on risk* (pp. 1–23). Earthscan.

Marais, K., et al. (2004, 29–31 March). *Beyond normal accidents and high reliability organizations: The need for an alternative approach to safety in complex systems* Engineering Systems Division Symposium, MIT, Cambridge, MA.

Marín-Puig, A., et al. (2022). Unattended gap in local adaptation plans: The quality of vulnerability knowledge in climate risk management. *Climate Risk Management, 38*, 100465. https://doi.org/10.1016/j.crm.2022.100465

Marschütz, B., et al. (2020). Local narratives of change as an entry point for building urban climate resilience. *Climate Risk Management, 28*, 100223. https://doi.org/10.1016/j.crm.2020.100223

McDaniels, T. (2021). Four decades of transformation in decision analytic practice for societal risk management. *Risk Analysis, 41*(3), 491–502. https://doi.org/10.1111/risa.13332

Mendelsohn, R. (2000). Efficient adaptation to climate change. *Climatic Change, 45*, 583–600.

Mudde, C. (2004). The populist zeitgeist. *Government and Opposition, 39*(4), 541–563.

Mudde, C. (2017). Populism: An ideational approach. In C. Rovira Kaltwasser, et al. (Eds.), *The Oxford handbook of populism* (pp. 28–47). Oxford University Press.

Nursey-Bray, M., & Palmer, R. (2018). Country, climate change adaptation and colonisation: Insights from an indigenous adaptation planning process, Australia. *Heliyon*, *4*(3), e00565. https://doi.org/10.1016/j.heliyon.2018.e00565

Oppenheimer, M., et al. (2019). *Discerning experts: The practices of scientific assessment for environmental policy*. The University of Chicago Press.

Oreskes, N. (2019). *Why trust science?* Princeton University Press.

Oreskes, N., & Conway, E. M. (2010). *Merchants of doubt: How a handful of scientists obscured the truth on issues from tobacco smoke to global warming*. Bloomsbury Press.

Ostrom, E. (1990). *Governing the commons: The evolution of institutions for collective action*. Cambridge University Press.

Paulson, S., et al. (2003). Locating the political in political ecology: An introduction. *Human Organization*, *62*(3), 205–217. http://www.jstor.org/stable/44127401

Perrow, C. (1984). *Normal accidents: Living with high-risk technologies*. Basic Books.

Perrow, C. (1999). *Normal accidents: Living with high-risk technologies* (2 ed.). Basic Books.

Perrow, C. (2011). Fukushima and the inevitability of accidents. *Bulletin of the Atomic Scientists*, *67*(6), 44–52. https://doi.org/10.1177/0096340211426395

Pescaroli, G., & Alexander, D. (2016). Critical infrastructure, panarchies and the vulnerability path of cascading disasters. *Natural Hazards*, *82*, 175–192. https://doi.org/10.1007/s11069-016-2186-3

Pidgeon, N., et al. (Eds.). (2003). *The social amplification of risk*. Cambridge University Press.

Reith, G. (2004). Uncertain times: The notion of 'risk' and the development of modernity. *Time & Society*, *13*(2/3), 383–402.

Rosa, E. A. (1998). Metatheoretical foundations for post-normal risk. *Journal of Risk Research*, *1*(1), 15–44. https://doi.org/10.1080/136698798377303

Scott, J., et al. (2000). Rational choice theory. In G. Browning (Ed.), *Understanding contemporary society* (pp. 126–138). SAGE.

Shepherd, T. G., et al. (2018). Storylines: An alternative approach to representing uncertainty in physical aspects of climate change. *Climatic Change*, *151*(3), 555–571. https://doi.org/10.1007/s10584-018-2317-9

Shrader-Frechette, K. S. (1991). *Risk and rationality: Philosophical foundations for populist reforms*. Univeristy of California Press.

Siegrist, M., et al. (2009). Salient value similarity, social trust and risk/benefit perception. In R. E. Löfstedt & Å. Boholm (Eds.), *The Earthscan reader on risk* (pp. 203–218). Earthscan.

Silvast, A., & Kelman, I. (2013). Is the normal accidents perspective falsifiable? *Disaster Prevention and Management, 22*(1), 7–16. https://doi.org/10.1108/09653561311301934

Slovic, P. (1992). Perception of risk: Reflections on the psychometric paradigm. In S. Krimsky & D. Golding (Eds.), *Social theories of risk* (pp. 117–152). Praeger.

Slovic, P. (2000). *The perception of risk.* Routledge.

Slovic, P., et al. (1986). The psychometric study of risk perception. In V. T. Covello, et al. (Eds.), *Risk evaluation and management* (pp. 3–24). Plenum.

Stripple, J., & Bulkeley, H. (Eds.). (2013). *Governing the climate: New approaches to rationality, power and politics.* Cambridge University Press. Cover image http://assets.cambridge.org/97811070/46269/cover/9781107046269.jpg

Villagrán de León, J. C., et al. (2012). Early warning principles and systems. In B. Wisner (Ed.), *The Routledge handbook of hazards and disaster risk reduction* (pp. 481–492). Rouledge.

Weick, K. E., et al. (1999). Organizing for high reliability: Processes of collective mindfulness. In R. L. Sutton & B. M. Straw (Eds.), *Research in organizational behaviour* (Vol. 21, pp. 81–123). Jai Press.

Zinn, J. O. (Ed.). (2008). *Social theories of risk and uncertainty: An introduction.* Blackwell.

CHAPTER 4

Cultural and Social Theories of Risk

Abstract This chapter addresses the counterforces of cultural and social risk theories to the technocratic and apolitical perspectives character-ising earlier conventional risk theory. Risk has, over the post-WW2 years, become commonplace in everyday language and media reports. At the same time, new theories of risk focusing on risks as socially constructed and imbued by culture have emerged. These risk theories acknowledge the importance of equity, politics and social stratification in understanding risk and the implications for risk assessment and management. Three such theories are discussed and analysed in this chapter: Cultural risk theory, Risk society theory and sustainability and climate change risk theory.

Keywords Anthropology · Constructivism · Cultural risk theories · Politics · Risk scholarship · Risk society · Sustainability risk theory

We refine, polish, and perfect our formal models for determining accept-able levels of risk, despite evidence that the assumptions and methods bear little relationship to the lay public's conceptions of problems. (Fiorino, 1989, pp. 503–504)

© The Author(s), under exclusive license to Springer Nature   65
Switzerland AG 2023
M. Granberg and L. Glover, *Climate Change as Societal Risk*,
https://doi.org/10.1007/978-3-031-43961-2_4

Over the post-war years, risk has become commonplace in everyday language and the media reports on risks with increased frequency (Lupton, 1999). It could be said that contemporary modern societies are obsessed with risk. But how risk is understood is changing, with the established scientific-technical approaches featuring numerically defined probability distributions being challenged by activists, researchers and by practical experiences (Fiorino, 1990). The Corona pandemic revealed how global and transboundary risks can occur in multiple social systems simultaneously and, as such, cannot be calculated as a linear function of probability and effects (Ringsmuth et al., 2022). Instead, certain risks must be contextualised in relation to institutionalised norms and the guiding values in society that frames both their understanding and handling (Choudhury, 2021). Therefore, risks are socially constructed, interpreted and narrated in their specific societal context (Krauß & Bremer, 2020). This approach, founded in an awareness of cultural context, challenges the former conventional understandings and definitions of risk, as discussed in the previous chapter, in ways that change the requirements for risk governance (Marschütz et al., 2020). These cultural and social perspectives also address the complicated challenge of creating coordinated, effective, impactful outcomes and priorities (Carr & Nalau, 2023).

Academic scholarship and its accompanying research programs are founded in disciplinary regimes (Thompson Klein, 1996), with different disciplines approaching risk studies differently (Althaus, 2005). In mathematics and statistics, it is a calculable problem; in economics, risk is considered a resource that can be utilised through calculations of probabilities; in medicine, it is a reality that can be controlled and managed, while in the social sciences, risk is viewed as a socially based, often subjective, but sometimes also with an interest in justice, politics and power (Kasperson & Kasperson, 2005).

Applying these knowledges in decision-making in most practice realms invariably challenges these disciplinary constraints. Recognising, understanding and responding to risk is a complex task and the more elaborate the systems involved and the greater their scale, the more difficult this endeavour becomes. No single scholarly discipline can span the range of this challenge. Further, as we move across the spectrum of hazard identification, risk assessment, risk management and risk communication, different scholarly disciplines are engaged sequentially in particular phases, with disciplinary knowledge being passed along, interpreted and incorporated into this chain of decision-making.

Hence, there are demands for interdisciplinary, multidisciplinary and transdisciplinary efforts to overcome the strictures of these arrangements (Hirsch Hadorn et al., 2008). Risk, therefore, is a broad theme attracting interest and scholarship from across the social and applied sciences. Here, we are interested in the contributions of risk themes and standpoints on risk from cultural and social perspectives (see Table 4.1).

Table 4.1 A selection of cultural and social risk themes and associated perspectives

Risk theme	Risk perspective
Evaluation Paradigms Theories and concepts that apply valuations and judgements of risk identification, responses and management	Environmental politics Governance institutions Social justice
Knowledge Types[a] Ways in which risk is understood; causal explanations	'Anti-science' views Indigenous/traditional ecological knowledge 'Post-normal' science Social science
Ideological Paradigms Political ideologies with specific normative goals and models of social relations, based on established political philosophies; risk theorising and conceptualisation	Environmental politics Social justice/Political economy
Risk Framings and Risk Perceptions Explanations of how/why perceptions of risk differ between stakeholders; the assignment of meaning; sources of interpretation	Cultural risk theory Environmental politics

[a]Knowledge types at a 'macro' level, as opposed to knowledge systems based in psychology, where categories such as conceptual, procedural, situational and strategic are employed

Other theories and conceptualisations could be added to this listing, such as governmentality and systems theory, but this selection provides a wide range of views prominent in risk scholarship, offering sharply varying perspectives and raises some of the associations between risk scholarship and broader social sciences debates (and, indeed, with those within society).

Of particular interest is the association between risk theories and politics, and the ways in which different risk understandings are translated into policies and practices. A number of different social science risk perspectives are relevant for understanding of climate change risks; below, three of the most prominent and relevant to the task of expounding the societal risk concept are reviewed: Cultural risk theory, Risk society theory and Sustainability and climate change risk theory, with a focus on their epistemological, ontological and political elements.

CULTURAL RISK THEORY

It has been long known—if not always necessarily analysed—that culture influences risk perception, reduction and management (Douglas & Wildavsky, 1983). Anthropologists and sociologists offer various cultural explanations of selected risks from the sorts of natural and technological hazards relevant to climate change (see, e.g., Crate, 2011). These approaches are at odds with scientific-technical descriptions of risk based on universal scientific principles, observations and deductions.

Viewing culture as a lens through which individuals and groups perceive risk is the 'social construction of risk' (Douglas, 1992). Within this perspective, one specific theory has become identified as its primary exemplar, the 'cultural theory of risk'. It is worth being mindful, however, that cultural theory (as it is known) is but one of many cultural or social risk constructions. At the outset, cultural theory should not be axiomatically taken as a counter-narrative to scientific-technical approaches. There are, however, sharp contrasts and incompatibilities between the two.

Cultural theory extends beyond narrow concerns with safety into the realms of political inquiry, debate and decision-making, dealing with intangible values and concepts, such as accountability, fairness, institutions, justice and representation (Lupton, 1999). Social and political variables are used in cultural theories of risk for explaining risk perception in corporations, groups, individuals and institutions. Culture, therefore, informs us as to what a social group perceives as risk, how this risk is assessed, how danger and disasters are understood, what measures are employed to determine the scale of risk and so on (Slovic, 2000). In its strongest form, cultural theory holds that risk perception can be explained by cultural factors and associated social learning (Thompson et al., 1990).

Cultural theory was, in part, a response to prominent contemporary debates over technology, prominently nuclear power and environmental pollution, with an interest in how public concern over this was shaped and changed over time. Douglas and Wildavsky (1983) describe risk as a *joint product* of knowledge of the future and social consent over the desired course of action. Critically, this includes recognising and evaluating risks to determine those considered dangerous. That the social milieu influences individual psychology in mundane and discrete decisions is an unremarkable claim, but cultural risk theories in general discern a more pervasive and systemic influence of culture (Clayton & Manning, 2018). Cultural phenomena, according to cultural risk theory, such as institutions, knowledge legitimation, norms, regulations and standards, exert a deep and powerful effect on individual perception used in a wide range of decisions concerning risk.

British anthropologist Mary Douglas is most closely associated with cultural explanations of risk in *Risk and Culture* (1983), usually regarded as the seminal text, co-authored with the American political scientist, Aaron Wildavsky. There is a universal tendency, they argue, for individuals to associate misfortune (i.e., social harms) with the transgression of certain norms. This results in social institutions being created that reinforce opposition to aberrant activity and act to regulate behaviour. Building on sociologist Emile Durkheim's functional institutional theory (Traugott, 1978), they develop the grid-group theory that theorises that society can be understood as expressing two themes lying along a continuum, one being that of social stratification that ranges from authoritarian and highly structured social life to the egalitarian, i.e., social regulation (the *grid* axis). Along the other thematic axis is the strength of collective control, with the highly regulated contrasting with the highly self-sufficient, i.e., social integration (the *group* axis). Society, therefore, can be depicted and understood as the extent of how social organisation is maintained through control over its individual members (i.e., regulation and stratification/ranking) and how its members are bound to a common cause or purpose (i.e., integration and the boundaries of the group). Set out as a two-by-two grid, four types result from this system (see Table 4.2).

Table 4.2   Cultural risk theory of grid-group typology

Fatalist (High Grid/Low Group) • Individuals apart from social groups, • Absence of control and order, and • Sceptical over risk prediction, risk management	Hierarchist (High Grid/High Group) • Individuals exist within hierarchical society, • Decision-making best done by elites, and • Risk management by elites and experts
Individualist (Low Grid/Low Group) • Disinterest in group membership; Individuals should have equal opportunities to maximise their own welfare and collective decisions/rules are not favoured, and • Risk management is an individual concern	Egalitarian (Low Grid/High Group) • Interest in social cooperation; group activity, • All are considered equal, and • Risk management aims for equitable outcome through equal treatment

Source After, Douglas and Wildavsky (1983)

Risk and Culture concentrates its attention on three types of organisations, a typology that uses the grid-group categories, to address the question of the social differentiation of attitudes towards risk and environmental disasters (Douglas & Wildavsky, 1983). Locating individuals within an organisational type is tied to their outlooks and beliefs. These cultural types are identified across different scales and can be applied to organisations, for this typology that could be said to 'personalise' institutional characteristics, as follows:

• *Individualist* organisations are 'low grid and low group' (i.e., weakly integrated and weakly regulated). These are market-oriented social groups. Risk management is the responsibility of individuals and not that of social groups (including governments). Necessary information and opportunities to take action should be universally available,

• *Egalitarian* organisations are 'low grid and high group' and form around environmental causes or issues and comprising a volunteer membership. Risk management involves arrangements that ensure that it produces equitable outcomes and aims for equal treatment. Central to the survival of such groups is the need to provide a narrative of the importance of their cause and magnifying the public presentation of environmental threats,

- *Hierarchist* organisations are 'high grid and high group' and risk management is the responsibility of elites in society; duties and responsibilities can be delegated to experts and managers and allocated according to hierarchical divisions of labour/responsibility, and
- *Fatalist* types are 'high grid and low group', where risk management is not of interest, as risks cannot be foreseen and therefore can be neither prevented nor ameliorated.

Great interest was attracted by the grid-group typology, and it is arguably the dominant means for analysing culture (notably in the administrative, organisational and political fields) and is widely applied to phenomena related to culture. It has been employed from the scale of individual types (i.e., as personality or behavioural types) all the way to cross-national comparisons (Spickard, 1989). This work is also the foundation of later models of attitudes towards Nature and the environment (i.e., Fatalist: Nature is capricious; hierarchical: Nature is perverse/tolerant; Individualist: Nature is benign, and Egalitarianist: Nature is ephemeral); these are also referred to as 'the myths of Nature' (see, e.g., Mamadouh, 1999).

Social construction has been prominent in scholarship on the social risks of climate change, with much use made of the cultural theory of risk, both for analysing individual and collective understandings of risk. As Jones (2011) observes of cultural theory, there is a 'remarkably consistent' link between the cultural theory types and attitudes towards climate change. 'Hierarchs' and 'egalitarians' both align with mainstream scientific opinion on climate change, albeit for different reasons, while individualists resist this science. In part, these responses are explained by the differences in how Nature is regarded, the legitimacy accorded to experts and attitudes to regulatory controls.

O'Riordan and Jordan (1999) find that the institutions developed to respond to climate change are based on 'contradictory' views of human behaviour, and use cultural theory to explain this condition. McNeeley and Lazrus (2014) use cultural theory and the 'myths of Nature' (excluding fatalism) to examine climate change adaptation in four case studies to explain how different worldviews influence the understanding of climate, finding that individualists view climate as naturally variable, hierarchicals view climate as manageable within limits and egalitarians view climate as having tipping points with potential of collapse.

Hulme (2009) reaches similar conclusions in *Why We Disagree about Climate Change* in considering how the different cultural types view climate change, based on the myths of Nature. Verweij et al. (2006) use the three positions to derive three 'stories' of climate change, each with its own setting, villain (i.e., policy problem), hero (i.e., policy protagonist) and moral (i.e., policy solution). To these can be added a considerable body of scholarship involving case studies from around the world. From these differing worldviews, it is possible to identify those risks of concern to each cultural theory type (see Table 4.3).

Table 4.3 Cultural risk theory types and climate change risks of greatest concern

Cultural theory type	Risks of greatest concern
Egalitarian	• Ethical and moral challenges to individuals and society • Increasing socio-economic inequity • Losses to ecosystem services • Degradation of environmental values • Inadequate governmental action, and • Corporate malfeasance/irresponsibility; exploitation of climate change by vested interests
Hierarchical	• 'Tragedy of the commons' ethos discourages effective responses • Inadequate government action and planning, especially at the global level • Failures of global/national climate agreements (covering formulation, contents, monitoring and enforcement) • Inadequate cooperation between civil society, the corporate sector and states, and • That expert advice is ignored/under-valued/ineffective
Individualistic	• That governments will impose punitive measures on society and corporations • That climate governance will enhance government powers in reach and strength • That climate governance will interfere with market adaptations to climatic changes, and • Markets will not be able to respond to climate change as appropriate signals are received

RISK SOCIETY THEORY

Risk society focuses on late modern society in which risk, and the responses to it, takes new and additional forms to those of earlier times and these changes shape society. Accordingly, this focus on risk indicates a transition from industrial society to risk society. Largely attributed to the sociologists Ulrich Beck (1992, 2005) and Anthony Giddens (1990, 2009), risk society is a grand social theory and is distinguished by its treatment of risk as a central element in the organisation of social life. Beck has a keen interest in the social production of scientific-technical risks and states that modernity and industrialisation has led to that condition where "…hazards and potential threats have been unleashed to an extent previously unknown" (Beck, 1992, p. 19). Initially, Giddens had a clearer focus on politics but in Beck's later work, the treatment of ecological themes is increasingly focused on politics and has much in common with concepts and themes raised in environmental politics and political economy. For Beck, the connections between risk, hazards and crises are linked to the construction and the capacity of political institutions (the political) and to politics (Beck & Cronin, 2006).

As a general guide, neither Beck nor Giddens typically feature in anthologies of key environmental thinkers or of key environmental writings. Within the field of sociology, however, these works have been highly influential, and Beck is regarded as the pre-eminent sociologist of the contemporary era. Beck's *Risk Society* (1992) covers natural and technological hazards but concentrates on the latter. Here, we draw largely on Beck's *Risk Society*; this thesis is covered in many other works by Beck, notably 'Risk society revisited: Theory, politics and research programmes' (1994), *World Risk Society* (1994) and *Ecological Politics in a Time of Risk* (2006).

In plain form, risk society is held to be that succeeding industrial society, a change in which risk is the central factor. Industrial society creates risks that are essentially bounded by space and time; they occur in singular locations and produce immediate hazards in that location, involving risks readily perceptible to the human senses. These risks are the consequences from inadequate, or incomplete, modernisation. A distinction is drawn between natural hazards and manufactured risks. In risk society, the character of risk changes dramatically; risks are those created by modernity and are unintended, cause and effect can be greatly distanced through space and time, and these risks are beyond perception through our primary senses.

As these risks become more widely appreciated and recognised, society becomes transformed from its previous industrial character; the next successive stage of modernity is the risk society. This social transition produces *reflexive modernisation*, according to Beck et al. (1994). Risk society is replete with special challenges. There are new and special risks to society created by human activity that can only be recognised and understood through contemporary knowledge (via technology). A further dimension to this conundrum is that much of this risk cannot be 'contained' in a finite sense. These risks, although they may be differentiated in their effects along socio-economic differences in society, have the feature of universality; both rich and poor alike are subject to them and wealth does not furnish an assured means of risk avoidance.

Beck's thesis is revolutionary; instead of society and its dynamic character being understood as being driven by the class, economic interests, gender, institutions, political power and other familiar variables, it has come to be shaped primarily by risk, i.e., it has become (or is becoming) a risk society (1992). Beck (1992, pp. 19–20) states:

> Questions of the development and employment of technologies (in the realms of nature, society and the personality) are being eclipsed by questions of the political and economic 'management' of the risks of actuality or potentially utilized technologies — discovering, administering, acknowledging, avoiding or concealing such hazards with respect to specially defined horizons or relevance.

Using nuclear accidents as an example, Beck states that normal understandings of risk are rendered obsolete by the duration, reach and scale of consequences of nuclear power disasters, which leaves these risks uncalculatable as "...the unknown and unintended consequences come to be a dominant force in history and society" (1992, p. 22). Further, public distrust in experts and state authorities has increased and this hampers state authorities in managing risks. This view is part of the wider change in recognising that distrust and scepticism cannot be reduced through increasing the provision of expert and scientific knowledge.

From an environmental perspective, the risk society thesis sits uneasily with many insights from environmental history. For a start, from the beginning of the Industrial Revolution there were risks that were displaced by space and time and beyond immediate perception of the senses (and indeed there are many examples from ancient history). Risk society theory holds that the divide between industrial society and risk society entails that of a change in risk perception. A problem here is that this transition has been underway for the entire duration of the industrial era; the notion of cause and effect being separate over time and/ or space has been a feature of industrial forms of pollution since it was first recognised. Risk society concerns the unintended consequences of modernity that exceed the capacities for control; this proposition places a considerable burden on the meaning of pollution 'control'.

Some environmentalists may reject the passivity of the claims that these risks were 'unintended' by pointing to deliberate efforts by corporations to overlook the consequences of their actions through wilful obfuscation, by withholding of evidence from the public and regulatory authorities, by states' failing to meet social responsibilities by investigation of possible or likely risks of compounds known to be dangerous, and so on (see, e.g., Oreskes & Conway, 2010). A further problem is *Risk Society's* distinction between natural hazards (as largely a pre-modern issue) and manufactured risks, especially in light of the losses from natural hazards in contemporary times. Furthermore, it is ironic that the role of culture is overlooked in considering the risks from natural hazards. *Risk Society* also appears to underplay the role of social factors in the distribution of risk, especially its effect in escalating the vulnerabilities of those with least resources.

Several scholars in environmental politics have studied Beck and risk society, usually within the 'ecological modernisation' framework (see, e.g., Hajer, 1995); by way of contrast, Beck's own work does not draw strongly on earlier environmental politics scholarship but on philosophical, political and sociological work, despite having several major works on environmental themes. Similarly, Anthony Giddens' *The Politics of Climate Change* (2009) received wide attention in the popular realm, yet its contribution, and its author's overall work on modernity, within the realm of environmental politics scholarship is much more unclear.

Harriet Bulkeley (2001), however, found aspects of climate change to 'epitomise' the risk society thesis. To begin, modernity is the root cause of (anthropogenic) climate change, so that the release of GHGs is an unintended hazard arising from modernity. Then there is the cause, effect and responsibility of climate change being stretched across space and time, transcending the limits of industrial risks, and whose effects will be experienced far into the future. Further, the risks of climate change can only be understood initially through science, another of the features of risk society.

SUSTAINABILITY AND CLIMATE CHANGE RISK THEORY

Although there is no formal theory of sustainability risk, there is a sizable volume of scholarship and research on the risks of climate change to environmental (and associated social) values and a large number of plans, policies and programs in business, civil society and government to address this risk that amounts to a sustainability risk discourse. Making this discourse unique is placing the environment at the centre of its concerns. Identified risks in this discourse are:

• Losses of environmental values, which include ecosystem services and natural resources; these include cultural, instrumental, material, service and spiritual values,
• Losses of ecological values, such as biodiversity, ecosystem processes, species and ecosystem distribution, and biotic interactions,
• Environmentally mediated social risks, such as occur through the spatial/temporal dislocation of hazards, and
• Interactions of climate change with other drivers of environmental loss, such as economic growth, population growth and urbanisation.

However, despite sharing a common theme in climate change, this risk discourse does not belong to any scholarly discipline, nor is it a discipline itself or sub-discipline—rather it spans the natural sciences, humanities and social sciences and other forms of knowledge, including traditional ecological knowledge. Furthermore, in addition to increasingly merging with the sustainability discourse, the climate change discourse now overlaps with that of disaster risk reduction (Birkmann & von Teichman, 2010).

Climate change threatens sustainability in three basic ways. Climate change: (1) Exacerbates, through direct and indirect impacts, existing conditions and trends of declining environmental values at all scales, from the local to the global, (2) Generates specific climate change impacts greater than those expected, and (3) Undermines the effectiveness of plans, policies and practices for improved sustainability.

As innumerable authors have argued, sustainability is widely and varyingly understood and without a common or accepted definition or meaning; consequently, it is a convenient, expedient, subjective and valuable term and concept in extensive use. In effect, this discourse combines several single discourses that always had some overlaps and that have merged over time, albeit the aspects of each contributor can still be distinguished. While this complex history provides rich pickings for scholars of the field, it has also led to confusion, questionable semantic debate and impediments in devising pragmatic measures. Rather than try to describe the range and tangled elements of the contemporary contents of his discourse, it is easier to begin with the two foundational themes, each of which has its own storyline, namely: (1) Nature conservation risks, and (2) Sustainable development risks; each is examined in turn.

Nature Conservation Risks

Arguably, a reverence for Nature is an ancient impulse and there are all manner of historical edicts for protecting species and natural resources, but the contemporary notions of Nature conservation are largely a product of industrial society's ambitions to safeguard (non-human) species and ecosystems from the harms rendered by society. In general, traditional hunter-gathering societies managed hunting practices and food harvesting to avoid depleting food species and their ecosystems. But it was Romantic literature, the developing biological sciences, species extinctions in Europe, European colonial expansions into the New World and changing cultural mores towards the natural world in the nineteenth century that underpinned the contemporary conservation movement (Evans, 2002). What follows is the steady expansion globally of conservation measures, typically produced by public policy, that includes declaring parks and reserves to secure species and ecosystems, protecting specific species and managing natural resource harvesting.

By the twentieth century, there were international efforts at Nature conservation, increasing notably following the UN's creation in 1945. Landmark developments include the 1948 establishment of the International Union for Conservation of Nature, followed by the World Wildlife Fund (est. 1961) and the United Nations Environment Programme (est. 1972). Advances in ecology, especially in the 1960s and 70s, and shaped by rising environmentalism, influenced the goals, priorities and practices in Nature conservation, with the 1980s development of conservation biology addressing the human causes of biodiversity loss and devising potential solutions (Soulé, 1985). Nature conservation expands outwards from its early reactive concerns from specific endangered species, natural resources and wilderness (and being 'crisis-driven') to include proactive decisions, anticipating and preventing potential conservation risks. Although Nature conservation originally addressed natural resource use, such as forestry and fishing, conservation biology had as its goals biodiversity preservation and restoration. As Robinson (2006, p. 663) observed, "People fall outside this sphere of interest..." because (excepting traditional societies) they are not 'natural' and are beyond the scope of analysis. Biodiversity is taken as being inevitably harmed by human activity.

While conservation biology may have been in a quandary over the place of society, there was strong use made of the human interests as a rationale for protecting biodiversity (such as being essential for human health and well-being, providing food security and making available genetic resources). There were various efforts to integrate Nature conservation with social goals by recognising the role and interests of society. In the *World Conservation Strategy* (IUCN, 1980), the three objectives are maintaining essential ecological processes and life support systems, using species and ecosystems sustainably and preserving genetic diversity. Indeed, the Strategy links economic development and Nature conservation, describing the lack of progress (notably in low-income nations) in the former as inhibiting that in the latter.

Biologists and ecologists recognise the risks of climate change for Nature conservation, especially for biodiversity, at the outset of the climate change discourse (see, e.g., IPCC, 1990). Climate change is already degrading Nature conservation values, especially in places of highest vulnerability, including coral reefs, mountainous areas and the polar regions. These losses include, obviously biodiversity loss, and reductions in species range, habitat destruction and changes to ecosystem functioning (see, e.g., IPCC, 2002).

Bringing the biodiversity and climate change responses together has lagged this awareness considerably (Pettorelli et al., 2021). There are, notably, separate UN conventions on climate change and biodiversity. Although the UN Convention on Biodiversity (United Nations, 1992) does not mention climate change, subsequent activities under the Convention identify climate change as a driver of change, such as by the UN body assessing the condition of biodiversity and ecosystem services (IPBES, 2019). There is also cognisance of climate risk inter-acting dynamically with social influences. Protecting biodiversity from further losses due to climate change is increasingly rationalised on poten-tial social benefits. These benefits are both highly numerous and diverse, including ecosystem services, for plant and animal health, for providing carbon stores (through sequestration) and for enhancing resilience to climate change impacts.

Sustainable Development Risks

One of the great and transformative social developments of the twentieth century was environmentalism and for many people, one of its key expres-sions is sustainable development (Schlosberg & Coles, 2016). Castro (2004) describes it as being contested, ubiquitous and indispensable, and few would disagree. As a great many works have been written on the disputations over its meaning and practices, and the controversies therein, under the necessity for brevity an overview of the sustainable develop-ment storyline is offered here to describe its identity and its unfolding relationship with the climate change discourse. Scholars have found many historical roots in the ideas about environmentalism and sustainable devel-opment (see, e.g., Du Pisani, 2006); here we confine our attention to the era of the climate change discourse.

Contemporary sustainable development is the marriage of two agendas marked by failure and disappointment by activists, policymakers and state and international institutions that becomes most apparent in the latter 1960s and early 1970s, namely the slow pace of economic progress in low-income nations under conventional economic development programs and the declining environmental conditions globally at the tail-end of the 'long boom'. These were the costs of a global industrial society: environmental destruction and inequality in distributing and sharing the costs and benefits of economic 'progress'. Expressions of these issues at the global, national and local levels abound during and before this time, with popular works from environmentalists (such as Paul Ehrlich's (1968) *Population Bomb*, Rachel Carson's (1962) *Silent Spring* and Barbara Ward and Rene Dubos's (1972) *Only One Earth*) and from critics of conventional development, such as Andre Gunther Frank (1966) and Paul Sweezy (1942).

Of the many ideas and concepts emerging out of these critiques was the recognising the linkages between economic development and environmental conditions, and that causal relationships could also serve as insights for designing and delivering remedial and mutually beneficial initiatives. Central to this link was the overriding goal that societal use of the environment should be sustainable—as should be the functioning of the economy, drawing on ecological models of species and ecosystems. Donella Meadows et al. (1972) *Limits to Growth* and Edward Goldsmith et al. (1972) *Blueprint for Survival* are influential in promoting limiting resource use and waste production to levels compatible with maintaining (or restoring) environmental conditions. Indicating the speed at which this idea spread was the 1972 *UN Conference of the Human Environment* in Stockholm, whose declaration seeks both economic development and environmental protection and promotes cooperation between developed and developing nations (United Nations, 1973). A key implication of limiting economic (and population) growth was that developing nations would not follow the same development trajectory as developed nations but would have to achieve their development 'sustainably' and that developed nations would have to engage in forms of large-scale transfers of finance, resources, technologies and the like to the developing world nations. Immediately apparent was that the environmental benefits of a new economic growth paradigm were not matched by the same certainty of meeting the development needs of developing nations.

If there is one event and document that sustainable development scholars refer to in sustainable development studies, it is the UN's World Commission on Environment and Development (WCED), and its major output, *Our Common Future* (WCED, 1987), known as the Brundtland Report. This report, in many respects, settled several major ambiguities in sustainable development. Famously, it furnished the most popular definition of sustainable development as "...development that meets the needs of the present without compromising the ability of future generations to meet their own needs" (WCED, 1987, p. 43); a definition that also highlights the risks of unsustainable development.

Although the report promulgates the tripartite themes of environment, economy and society (to later become 'triple bottom line' reporting), it strongly promotes sustainable economic growth, holding that such growth alleviates poverty and that poverty itself was a major driver of vulnerability and environmental losses in developing nations. Sustainable development is nothing less than an attempt to redefine 'progress'.

Sustainable development has been highly successful with widespread and global adoption by business, civil society and governments, with a virtually uncountable usage in policies and plans. As Sachs (2012, p. 2206) observes:

> Almost all the world's societies acknowledge that they aim for a combination of economic development, environmental sustainability, and social inclusion, but the specific objectives differ globally, between and within societies.

While it is not true that sustainable development offers to be all things to all people, the concept has succeeded in becoming the central focus for a socially progressive form of progress. Robinson (2004) summarised the criticisms of sustainable development as being its vagueness, attractiveness to hypocrites and engendering of delusions of being effective; although many more have been made (see, e.g., Redclift, 1987). An examination of the UN's 17 Sustainable Development Goals (United Nations, 2015) is indicative of this ambition, as these now include goals on education, gender equity and peace, justice and strong institutions, along with earlier themes with such goals as those pertaining to economic growth, health, hunger and poverty. Hence, sustainable development now casts a very wide net.

As the climate change discourse has grown and intensified, it has influenced the sustainable development discourse (Beg et al., 2002). A study of major international policy and program documents in sustainable development finds numerous links to the risks of climate change, such as the UN Sustainable Development Agenda. Climate change impacts directly worsen the prospects for nearly all goals sustainable development goals. As Downing et al. (2003, p. 3) suggest, "...climate change and sustainable development interact in a circular fashion". Critically, it was also realised that development pathways influence: (1) Future GHG emissions through such decisions as energy pathways, and (2) Responses to climate change impacts, influence adaptations, resilience, transformation potentials and vulnerabilities. Consequently, there are calls for greater integration between sustainable development and climate change (see, e.g., Swart et al., 2003).

At the same time, the disaster risk discourse has also linked with those of climate change and sustainable development. This connection is shaping these discourses. For example, the UN Office for Disaster

Risk Reduction's (UNDRR) assessment of disaster risk observed that the climate change discourse is moving from a static view of risk framing to "…a more dynamic framing where responses to the risks with potential side effects and interactions among risks are more strongly considered" (UNDRR, 2022, p. 5). Little explanation is needed for this discourse linkage, as over recent decades, meteorological extreme events have increased (WMO, 2021), with climate change designated as the cause, a trend that will continue (IPCC, 2022b). Growing population and urbanisation result in more communities and infrastructure at risk from climate hazards, whose intensity and frequency are increasing under climate change (IPCC, 2022a; Van Aalst, 2006) and the scale of these hazards is shown in the trends of increasing mortality and morbidity from these climatic hazards (see, e.g., WMO, 2021). Practitioners in the climate change and disaster risk fields have turned to the sustainable development discourse to better understand the interactions between social and environmental factors shaping hazards and as a template for devising and implementing response measures.

References

Althaus, C. E. (2005). A disciplinary perspective on the epistemological status of risk. *Risk Analysis, 25*(3), 567–588. https://doi.org/10.1111/j.1539-6924. 2005.00625.x

Beck, U. (1992). *Risk society: Towards a new modernity*. Sage.

Beck, U. (2005). *Power in the global age: A new global political economy*. Polity.

Beck, U. (2006). Risk society revisited: Theory, politics and research programmes. In J. F. Cosgrove (Ed.), *The sociology of risk and gambling reader* (pp. 61–83). Routledge.

Beck, U., & Cronin, C. (2006). *Cosmopolitan vision*. Polity.

Beck, U., et al. (1994). *Reflexive modernization: Politics, tradition and aesthetics in the modern social order*. Polity Press.

Beg, N., et al. (2002). Linkages between climate change and sustainable development. *Climate Policy, 2*(2–3), 129–144. https://doi.org/10.3763/cpol. 2002.0216

Birkmann, J., & von Teichman, K. (2010). Integrating disaster risk reduction and climate change adaptation: Key challenge—Scales, knowledge, and norms. *Sustainability Science, 5*(2), 171–184.

Bulkeley, H. (2001). Governing climate change: The politics of risk society? *Transactions of the Institute of British Geographers, 26*(4), 430–447. https:// doi.org/10.1111/1475-5661.00033

Carr, E. R., & Nalau, J. (2023). Adaptation rationales and benefits: A foundation for understanding adaptation impact. *Climate Risk Management, 39*, 100479. https://doi.org/10.1016/j.crm.2023.100479

Carson, R. (1962). *Silen spring*. Houghton Mifflin Company.

Castro, C. J. (2004). Sustainable development: Mainstream and critical perspectives. *Organization & Environment, 17*(2), 195–225.

Choudhury, B. (2021). Climate change as systemic risk. *Berkeley Business Law Journal, 18*(2), 52–93.

Clayton, S., & Manning, C. (Eds.). (2018). *Psychology and climate change: Human perceptions, impacts, and responses*. Academic Press.

Crate, S. A. (2011). Climate and culture: Anthropology in the era of contemporary climate change. *Annual Review of Anthropology, 40*(1), 175–194. https://doi.org/10.1146/annurev.anthro.012809.104925

Douglas, M. (1992). *Risk and blame: Essays in cultural theory*. Routledge.

Douglas, M., & Wildavsky, A. (1983). *Risk and culture: An essay on the selection of technological and environmental dangers*. University of California Press.

Downing, T. E., et al. (2003). Special supplement on climate change and sustainable development. *Climate Policy, 3*, 3–8.

Du Pisani, J. A. (2006). Sustainable development–historical roots of the concept. *Environmental Sciences, 3*(2), 83–96.

Ehrlich, P. R. (1968). *The population bomb*. Sierra Club.

Evans, D. (2002). *A history of nature conservation in Britain*. Routledge.

Fiorino, D. J. (1989). Environmental risk and democratic process: Critical review. *Columbia Journal of Environmental Law, 14*(2), 501–548.

Fiorino, D. J. (1990). Citizen participation and environmental risk: A survey of institutional mechanisms. *Science, Technology, & Human Values, 15*(2), 226–243. http://www.jstor.org/stable/689860; https://www.tandfonline.com/doi/full/10.1179/2046905514Y.0000000146

Frank, A. G. (1966). The development of underdevelopment. *Monthly Review, 18*(4), 17–31.

Giddens, A. (1990). *The consequences of modernity*. Stanford University Press.

Giddens, A. (2009). *The politics of climate change*. Polity Press.

Goldsmith, E., et al. (1972). *A blueprint for survival*. Houghton Mifflin.

Hajer, M. (1995). *The politics of environmental discourse: Ecological modernisation and the policy process*. Oxford University Press.

Hirsch Hadorn, G., et al. (Eds.). (2008). *Handbook of transdisciplinary research*. Springer.

Hulme, M. (2009). *Why we disagree about climate change: Understanding controversy, inaction and opportunity*. Cambridge University Press.

IPBES. (2019). *Summary for policymakers of the global assessment report on biodiversity and ecosystem services of the intergovernmental science-policy platform on biodiversity and ecosystem services*. Intergovernmental Science-Policy Platform of Biodiversity and Ecosystem Services (IPBES).

IPCC. (1990). *Climate change: The IPCC impacts assessment*. Intergovernmental Panel on Climate Change (IPCC).

IPCC. (2002). *Climate change and biodiversity*. Intergovernmental Panel on Climate Change (IPCC).

IPCC. (2022a). *Climate change 2022: Impacts, adaptation and vulnerability*. Intergovernmental Panel on Climate Change (IPCC).

IPCC. (2022b). *Climate change 2022: Mitigation of climate change*. Intergovernmental Panel on Climate Change (IPCC).

IUCN. (1980). *World conservation strategy*. International Union for Conservation of Nature and Natural Resources (IUCN).

Jones, M. D. (2011). Leading the way to compromise? Cultural theory and climate change opinion. *PS: Political Science & Politics, 44*(4), 720–725. https://doi.org/10.1017/S104909651100134X

Kasperson, J. X., & Kasperson, R. E. (Eds.). (2005). *The social contours of risk* (Vol. I). Earthscan.

Krauß, W., & Bremer, S. (2020). The role of place-based narratives of change in climate risk governance. *Climate Risk Management, 28*, 100221. https://doi.org/10.1016/j.crm.2020.100221

Lupton, D. (Ed.). (1999). *Risk and sociocultural theory* (2nd ed.). Cambridge University Press.

Mamadouh, V. (1999). Grid-group cultural theory: An introduction. *GeoJournal, 47*(3), 395–409. https://doi.org/10.1023/A:1007024008646

Marschütz, B., et al. (2020). Local narratives of change as an entry point for building urban climate resilience. *Climate Risk Management, 28*, 100223. https://doi.org/10.1016/j.crm.2020.100223

McNeeley, S. M., & Lazrus, H. (2014). The cultural theory of risk for climate change adaptation. *Weather, Climate and Society, 6*(4), 506–519. https://doi.org/10.1175/WCAS-D-13-00027.1

Meadows, D. H., et al. (1972). *The limits to growth*. Universe Books.

O'Riordan, T., & Jordan, A. (1999). Institutions, climate change and cultural theory: Towards a common analytical framework. *Global Environmental Change, 9*(2), 81–93. https://doi.org/10.1016/S0959-3780(98)00030-2

Oreskes, N., & Conway, E. M. (2010). *Merchants of doubt: How a handful of scientists obscured the truth on issues from tobacco smoke to global warming*. Bloomsbury Press.

Pettorelli, N., et al. (2021). Time to integrate global climate change and biodiversity science-policy agendas. *Journal of Applied Ecology, 58*(11), 2384–2393. https://doi.org/10.1111/1365-2664.13985

Redclift, M. (1987). *Sustainable development: Exploring the contradictions*. Methuen.

Ringsmuth, A. K., et al. (2022). Lessons from Covid-19 for managing transboundary climate risks and building resilience. *Climate Risk Management*, 35, 100395. https://doi.org/10.1016/j.crm.2022.100395

Robinson, J. (2004). Squaring the circle? Some thoughts on the idea of sustainable development. *Ecological Economics*, 48(4), 369–384.

Robinson, J. G. (2006). Conservation biology and real-world conservation. *Conservation Biology*, 20(3), 658–669. https://doi.org/10.1111/j.1523-1739.2006.00469.x

Sachs, J. D. (2012). From millennium development goals to sustainable development goals. *The Lancet*, 379(9832), 2206–2211.

Schlosberg, D., & Coles, R. (2016). The new environmentalism of everyday life: Sustainability, material flows and movements. *Contemporary Political Theory*, 15(2), 160–181. https://doi.org/10.1057/cpt.2015.34

Slovic, P. (2000). *The perception of risk*. Routledge.

Soulé, M. E. (1985). What is conservation biology? *BioScience*, 35(11), 727–734. https://doi.org/10.2307/1310054

Spickard, J. V. (1989). A guide to Mary Douglas's three versions of grid/group theory. *Sociology of Religion*, 50(2), 151–170. https://doi.org/10.2307/371 0986

Swart, R., et al. (2003). Climate change and sustainable development: Expanding the options. *Climate Policy*, 3(Suppl. 1), 19–40.

Sweezy, P. M. (1942). *The theory of capitalist development*. Monthly Review Press.

Thompson Klein, J. (1996). *Crossing boundaries/knowledge, disciplinarities, and interdisciplinarities*. University of Virginia Press.

Thompson, M., et al. (1990). *Cultural theory*. Westview.

Traugott, M. (Ed.). (1978). *Emile Durkheim on institutional analysis*. The University of Chicago Press.

UNDRR. (2022). *Global assesment report on disaster risk reduction 2022*. United Nations Office for Disaster Risk Reduction (UNDRR).

United Nations. (1973). *UN conference of the human environment*. United Nations.

United Nations. (1992). *Convention on biodiversity*. United Nations.

United Nations. (2015). *Transforming our world: The 2030 agenda for sustainable development*. United Nations.

Van Aalst, M. K. (2006). The impacts of climate change on the risk of natural disasters. *Disasters*, 30(1), 5–18.

Verweij, M., et al. (2006). Clumsy solutions for a complex world: The case of climate change. *Public Administration*, 84(4), 817–843. https://doi.org/10.1111/j.1540-8159.2005.09566.x-il

Ward, B., & Dubos, R. (1972). *Only one earth*. Norton.

WCED. (1987). *Our common future*. Oxford University Press.

WMO. (2021). *WMO atlas of mortality and economic losses from weather, climate and water extremes (1970–2019)*. World Meteorological Organization (WMO).

PART III

Collapse and Transformation

CHAPTER 5

Understanding Climate Change as Societal Risk

Abstract Societal risk is advanced as a concept that is differentiated from the generic social risks of climate change. Societal risk is formally defined and its four constituent elements are expounded: (1) Ecosystem services cover provisioning, regulating and supporting services and are essential for human life; all are vulnerable to climate change and have been degraded through inadequate maintenance, pollution, overuse and other factors, (2) Cultural entities and processes are equally tied to climatic factors and their attendant benefits, such as, are at risk of change, degradation or loss: health, migration and security risks are reviewed, (3) Although often intangible, non-material cultural/social entities at risk are essential to societies and include aesthetic, artistic and heritage benefits and (4) Ecological values are being eroded by societies; society has an ethical responsibility and material necessity to recognise and address the climate risks to ecology.

Keywords Ecosystem services · Ecological values · Health · Migration · Security · Societal risk · Socio-ecological services · Values

© The Author(s), under exclusive license to Springer Nature   89
Switzerland AG 2023
M. Granberg and L. Glover, *Climate Change as Societal Risk*,
https://doi.org/10.1007/978-3-031-43961-2_5

COMPONENTS OF SOCIETAL RISK

Societal risk can be distinguished from social risk and there are good reasons for doing so, especially regarding climate change. Societal risk can be defined and understood in more precise and clearer ways than the generic concept of social risk; it can also incorporate an understanding of society-Nature relations and be used to promulgate climate justice goals. It comprises four elements (see Table 5.1):

- *Socio-ecological systems and outputs*: Human ecology from the applied sciences provides the reasoning for identifying the essential socio-ecological goods and services on which all societies ultimately depend, such that the loss (or threat of) diminution of one or more these services, either temporarily or permanently, imperils a society,
- *Social/cultural entities and processes*: Social sciences provide the reasoning for identifying the essential social and cultural elements that bond, identify, legitimate and sustain social units. Although some scholars view such social dimensions of human survival as being secondary effects to the external influences of changes to ecosystem, it is argued here that these factors can have a degree of independence from ecosystem services and are, of themselves, a component of societal risk,

Table 5.1 Components of societal risk

Component	Source of values/Benefits	Character of values/Benefits
Socio-ecological systems and outputs	Ecological goods and services as exploited by social activities	Material
Social/Cultural entities and processes	Social institutions, civic society, governments, businesses	Material
Non-material social/Cultural entities and processes	Ecosystem services, ecosystem processes and entities, cultural processes	Non-material
Ecological values	Ecology	Intrinsic

- *Non-material social/cultural entities and processes*: Social sciences also provide the reasoning that key non-material values are an additional element of society, such as collective social values, cultural elements and spiritual and religious views, and can be both at risk and essential to the continuation of social functioning and existing social structures. Hazards to this element are more likely to be indirect, or consequential, impacts of external hazards. Place is critical to societies' lived experience; climate change threatens this association, as do some adaptations, such as migration, and
- *Ecological values*: Ecology and related natural sciences provide an account of the risks to non-human species, ecosystems and ecological processes at risk from anthropogenic influences; these entities have *intrinsic* value over which societies have responsibility.

Below, the climate change societal risks are described for these components. Both objective and subjective factors are employed; the former including such elements as the duration, extent, intensity, persistence, scale and timing of the climate change impacts on these components, and the latter including equity and distribution factors, and intrinsic valuations.

On the Meaning of 'Values'

Reviews of the risk concept (in preceding chapters), and its applications to climate change hazards, reveal the central place of values. However, as Tschakert et al. (2017) point out, 'values' has several different meanings and applications in the climate change discourse:

- Economically, such as shown by markets (as use, or non-use, values),
- Ecologically, where valuations of Nature can be instrumental or intrinsic (see below),
- Philosophically, where values are an aspect of ethics and moral principles,

- Psychologically, such as when used in concepts such as universal human values, and
- In everyday use, where values refer simply to that which is valued and at risk of loss.

In considering how to define, understand and apply the societal risk concept, these differing conceptualisations of values are evoked as the contributions of different fields of knowledge are applied, as shown below.

CLIMATE RISKS TO SOCIO-ECOLOGICAL SYSTEMS AND OUTPUTS

Appreciating the importance of socio-ecological systems to human survival, whether for individuals or entire nations, is simultaneously obvious and immensely difficult to fully appreciate. Society derives benefits from ecosystems through transformation processes that convert ecosystem resources into flows of goods and services essential to human life and well-being (and prosperity), i.e., the *ecosystem services* (see, e.g., Costanza et al., 1997). Well-being is a subjective quality influenced by culture, geography and other factors but is generally taken as being a life characterised by basic material needs being met, with various freedoms, health and security as furnished within society (noting that some include happiness as a well-being component); here, we are concerned with material needs for societal well-being. Ecosystem services supply tangible goods and services derived from natural and modified ecosystems (as socio-ecological systems) comprising: (1) Provisioning services, (2) Regulating services, and (3) Supporting services (see Table 5.2). This typology is based on the UN's Millennium Ecosystem Assessment (MEA, 2005) with one important modification, that being that the treatment here of cultural services and intangible benefits as separate components of societal risk (see Table 5.2).

Table 5.2 Types of ecosystem services and associated risks

Ecosystem service	Characteristic	Exemplars of benefits/Services at risk
Provisioning	Provision of materials and services	Biochemical Energy/Fuel Fibre Food/Nutrition Genetic material Medicines Water supply
Regulating	Benefits provided by regulation of ecosystem processes	Carbon sequestration Climate regulation Disease regulation Erosion control Pollination Water purification Water flow regulation
Supporting	Services enabling/underpinning the productivity of other ecosystem services	Primary production Nutrient cycling Soil formation

Source After, MEA (2005)

Ecosystem services are essential for human survival and whose diminution or loss necessarily generates a cost that ranges from minor difficulties to perilous conditions forcing migration or, in extreme cases, contributing to the collapse of a civilisation (Dobson et al., 2006). While there are universal human needs for survival, the use of ecosystem services and the conduct of socio-ecological systems is essentially as varied as are societies, each using and managing these systems depending on place and according to culture and its attendant governance, institutions and technologies. Interactions between ecosystem services vary considerably, from the independent to the tightly linked. Furthermore, oftentimes the climate–society relationship is a dynamic one, adding to the complexity of ecosystem service use. Climate change presents a difficult problem for socio-ecological systems, as it creates impacts across a span of interacting ecological elements; climate change impacts often compound social existing stressors and drivers of change (such as changes in land use and land cover, species introduction, technology use, resource consumption and waste production) and physical and biological drivers (IPCC, 2022; MEA, 2005).

From a societal perspective, climate change is a crisis heaped upon the many existing crises in the ecosystem services vital to society. Increasing production of desired goods and services, such as energy, food, fibre, timber and water causes declining productivity and viability in these and related ecosystem components (see Box 5.1). Human activity has taken a toll on the natural environment and this, in turn, has lessened and weakened those vital ecosystem services. Essentially, two processes have created this dilemma, that of excessive resource consumption and of pollution outputs, the net global trends of which are worsening (UN Environment, 2019).

Only a few indicators are needed to illustrate the seriousness of this problem. These ecosystem services sustain social life but have reached, at the global level, an unsustainable scale of harvesting (Arrow et al., 1995). Vitousek et al. (1986) estimated that 40% of the net global primary productivity of terrestrial ecosystems was co-opted by social systems annually. Smil (2016) reckoned that, in 2015, the weight of the world's domesticated farm animals was at least 25 times that of all wild mammals and that the living weight of 7.3 billion people was second only to that of domesticated cattle, making ourselves and cattle the dominant vertebrate species. Using the ecological footprint model to estimate the world's resource use relative to its regenerative capacity, the Global Footprint Network estimated that in 2014 (using 2018 data), we used 1.7 worth of the earth's resources (Lin et al., 2018).

Climate change has become a major driver of the decline of ecosystem services and serves to multiply the effects of other drivers. This transformation occurs through socio-ecological systems that interconnect ecological and social systems, for providing societal well-being depends on those social systems and the related socio-ecological services.

Box 5.1   Background: a snapshot of socio-ecological systems losses

---

- 85% of wetlands area have been lost,
- Land productivity has declined in 23% of land area,
- 75% of the land surface is significantly altered,
- One-third of the land surface is grazed or cropped, and
- Of 6190 domesticated plants and animals, over 1000 are threatened, 559 have become extinct

---

Source IPBES (2019)

CLIMATE RISKS TO SOCIAL/
CULTURAL ENTITIES AND PROCESSES

No society, from the traditional indigenous peoples drawing on local resources using traditional ecological knowledge in developing nations to city-dwellers in developed nations enmeshed in the global economy, can escape the impacts of climate change on their culture (Adger et al., 2013). Culture, always a fraught concept, is taken to be the beliefs, customs, symbolic practices and values shaping collective social behaviours and perceptions of an identified social group. Although some scholars take a different view, here culture includes its enablers in technology and institutions, and its material outputs. Societies form, exist, evolve, flourish, (and decline) in a climatic context, the influence of which is highly variable.

Changes to climate-sensitive ecosystem services, such as those that cease to exist, are diminished, change in duration, extent, intensity or timing or have their interactions with other systems are altered, are likely to have implications for the relevant socio-ecological services and, in turn, for social and cultural entities and practices (see Table 5.3). Social systems subject to climate influences, especially those related to social institutions, are also affected by climate change. Societal cohesion, composition, identity and sense of place are bound up with climatic influences, as are material (and non-material) dimensions of culture. As culture is central to our understanding of climate change, shaping ethics, perceptions and values, and playing a major role in determining the defining characteristics of any society, climate change thereby alters our views of climate and of society.

Table 5.3 Types of social and cultural entities and processes at risk

Societal risk component	Source of benefits	Exemplars of benefits at risk
Social and cultural entities and processes	Direct and indirect tangible outputs from socio-ecological systems and from social systems influenced by climatic (and related) factors	Education Health Mobility/Migration Recreation Security

Of the numerous risk themes in this area, three are of particular interest and have drawn wide attention, namely the risks to health, migration of displaced peoples and security, all of which are significant at the societal level.

Health

That climate change was a major risk to human health on a global scale was recognised early in the climate change discourse (see, e.g., McMichael et al., 2003) and there has been considerable research into the ongoing and forecast effects on mortality and morbidity (see, e.g., IPCC, 2023). Direct effects of climate on health can occur from extreme climate events and indirectly from impacts on ecosystems, on socio-ecological systems or on the social and economic realms, demonstrating the often-complex character of the climate–health relationship. Where multiple health risks are coincidental, there are compounding health risks. Climate change health risks result from increasing existing health hazards, namely a widening and deepening of health vulnerabilities combined with expansions of exposed populations; it is the additional health risks to already vulnerable peoples that are of greatest concern.

McMichael et al. (2006) lay out the main pathways that climate change affects population health, identifying major health effects from:

- *Extreme weather events*: whose impacts include effects on food yields, injury/death from natural disasters and thermal stress,
- *Effects on ecosystems and particular species*: whose impacts include changes in vector–pathogen–host relations, changes in infectious disease geography/seasonality, food poisoning, microbial proliferation and unsafe drinking water,
- *Sea-level rise, salination and storm surge*: whose impacts include reduced crop, fisheries and livestock yields, leading to impaired health, nutrition and survival (this also applies to the effects on ecosystems, as above), and
- *Ecosystem degradation*: whose impacts include displacement, leading to poverty and adverse health outcomes (e.g., Infectious diseases, malnutrition, mental health and physical risks) and loss of livelihoods.

Climate-sensitive diseases exact a great toll on human health; globally, deaths from these diseases are estimated at over 39 million for 2019 (Vos et al., 2020). Although many health conditions are related to climatic factors, as assessed by attribution to global deaths, a few stand out over recent decades (in order of descending importance): Cardiovascular diseases, death from malignant neoplasms, non-communicable respiratory illness, respiratory tract infections, Diabetes, diarrhoeal diseases, Malaria, nutritional deficiencies, Salmonella, environmental heat and cold exposure and Dengue (noting that matching data on mental health is unavailable) (IPCC, 2022).

Injuries resulting from extreme weather events are significant on a global scale, but their greatest occurrence is at the event sites and surrounds. Some 396 extreme weather and climate events in 2019 killed 11,755 people, affected 95 million at a cost of around USD130 million, the majority from floods and storms (Ebi et al., 2021). Inter-annual variability for extreme events is high and therefore, so are the associated risks. Eckstein et al. (2017) found global fatalities from such events for the period 1998–2017 to be 526,000 arising from 11,500 events.

Distinguishing the effect of climate change on health from non-climatic factors makes the climate risk assessment difficult, but there are clear trends that climatic change is already producing worsening health outcomes. Extreme events are forecast to increase (in extent, frequency and intensity), thereby increasing the risks of death and injury. Climate-sensitive health risks also increase. A WHO (2018) report presented research findings that projected that, because of climate change:

- Up to three billion people aged over 65 could be subject to heatwaves by 2100,
- By 2030, 100 million people could be forced into poverty, and
- 250,000 additional annual deaths could occur between 2030 and 2050 from four main health impacts.

Increases in communicative and non-communicative diseases include a shift in Malaria to higher altitudes and an increase in Dengue, Chikungunya virus, Lyme disease and tick-borne encephalitis (IPCC, 2022). Higher temperatures and rainfall are associated with more numerous diarrhoeal and gastrointestinal diseases cases, including Cholera and in water-borne diseases.

A somewhat neglected health issue is that of mental health risks. These cover a cluster of risks, including to direct mental health (e.g., depression, PTSD and suicide), diminished well-being (e.g., stress and cognitive impairment) and weakened social relations (e.g., loss of culture and interpersonal violence). Cunsolo and Ellis (2018) refer to *ecological grief* to describe the emotional and mental response by those who experience losses due to climate change (aka *ecoanxiety, ecogrief* and *climate trauma*). Cyclones, fires and floods, among other natural hazards, have been shown to produce lasting effects on mental health and can have community-wide effects. Berry et al. (2010), for example, describe how in that, in communities, reduced physical health and the trauma of events are related pathways to negative mental health impacts arising from climate change-related events. These authors also point to the differences between direct and indirect effects on mental health arising from 'acute' weather events and 'chronic' climatic events.

Migration

In contemporary literature, there is a concept describing places of societal habitation as a *human climate niche*, that "…is shaped by direct effects of climate on us and indirect effects on the species and resources that sustain or afflict us" (Lenton et al., 2023, p. 2). As conditions deteriorate in some areas and improve in others, adaptation in-place might not be possible in all locations, driving the need for migration to more livable climate niches (Xu et al., 2020).

Migration resulting from climate change impacts attracts great interest from policymakers and researchers, but remains highly controversial (Kniveton et al., 2008). What is of interest for policymakers and government officials is, of course, mass migration—the prospect of which has at times evoked a near-hysterical response in the mass media over the prospect of Europe and North America being overwhelmed by *climate refugees*—although exactly what qualifies as mass migration is open to question. Certainly, there is widespread concern over climate-related migration as climate change worsens in the coming decades, especially in light of its development and humanitarian implications for many of the world's poorest nations. A World Bank report (Clement et al., 2021) forecast that by 2050, without significant preventative actions, six regions of the world would see over 216 million people move within their nations due to (slow onset) climate change, centring on migration 'hotspots'.

Sub-Saharan Africa, East Asia and the Pacific and South Asia are the most afflicted.

Climate change intersects with the economic, political and social factors acting to force migration or to attract migrants; coincidental with these are the institutional, legal and political factors determining national and regional responses to migration. Migration produces changes from immigration for the host location *and* those resulting from emigration for the source location. Climate change overlays on the existing elements influencing migration, so that it may prompt new migration or promote an existing migration pattern, but it can also decrease existing migration. In short, climate change can influence not only the locations and scale of migration, but also its extent and duration. In the language of climate change adaptation, migration is the equivalent of abandonment on a temporary or permanent basis and occurs when coping capacity is exceeded.

Migration assumes many forms:

- Displacement, occurring after extreme events and may be temporary or permanent,
- Voluntary migration (or adaptive migration), where the migration decision is relatively unencumbered and made by individuals/families/households,
- Involuntary migration, where there is little choice over the matter,
- Relocation (or planned/organised resettlement), where migration is centrally organised for a population group,
- Rural-to-urban migration, that can be temporary or permanent, and
- Immobility, where there is an inability/unwillingness to migrate despite reasons to do so.

Although most attention is given to cross-border migration with its political and humanitarian implications, most climate-related migration occurs within nations, i.e., *internal* migration. Temporary or seasonal migration (such as transhumance by nomadic pastoralists) is widespread and typically occurring in tune with the seasons. This multiplicity of migration forms has led some to reject the term 'migration' and promote the concept of *climate mobilities* to better capture this range of responses (see, Boas et al., 2019).

Sudden-onset disasters comprise a direct climate impact and contrast with slower-onset disasters evoking indirect impacts of climate change. In the climate context, migration is a response to deleterious environmental change, the degradation or loss of environmental services or related factors undermining local or regional economies in combination with other socio-political factors. Extreme weather events can also produce lasting socio-economic effects contributing to migratory pressures over time, especially in cases where the frequency/intensity of these events is increasing, and full post-disaster recovery is impossible to achieve.

To date, climate-related migration has been predominantly a rural matter involving those making a living from agriculture and natural resource harvesting, and confined within the borders of lower- and middle-income nation states. Where there is cross-border migration, this usually occurs as part of rural migration where regions span nation state borders, although this is not a common occurrence. Yet, for all the alarmism over climate refugees entering Europe and US, there is no supporting evidence of this occurring (IPCC, 2022). Indeed, of the available research using empirical data, there are conflicting findings over the climate-migration link from different localities. What is far clearer, however, is the scale of population displacement following extreme climate events (notably extreme storms and floods). Although these annual displacement tallies vary considerably year-to-year, IPCC (2022) estimate the annual level worldwide exceeds 20 million.

Migration is a difficult issue to predict because it is considerably affected by local circumstances, such that explanatory factors can vary greatly between locations and communities and, of course, because the key aspects of climate change impacts are so uncertain, especially at the local scale. Although the poor are typically more likely to migrate, there are circumstances where a community's poorest are unable or unwilling to migrate. It is possible that slow onset changes (i.e., *chronic climate change*) will induce greater migration than quick onset changes (i.e., *acute climate change*), with most activity concentrated in rural, urban and coastal systems (see, Clement et al., 2021).

As climate change impacts continue to accumulate, many societies face no other prospect than migration and, in most cases, this will be a permanent abandonment of their home locations because continued habitation will no longer be possible or safe (Kniveton et al., 2008). Those unable to migrate, and this will be disproportionally society's poorest, women, children and the elderly, will be left in the world's most vulnerable locations and are likely to have the least adaptive capacities. Immobility can arise from many causes, ranging from being physically or financially unable to migrate, to those whose ties to where they live are stronger motivations than the reasons for leaving.

Security

Climate change is increasingly seen as an international security issue by scholars, think tanks and international agencies, including the UN Security Council (Maertens, 2022). Nordås and Gleditsch (2015) find that studies of the security implications of climate change are inconclusive and contrast with policy debates centring on claims of greatly increasing and violent conflicts. Sellers et al. (2019), on the other hand, identify several pathways linking the health losses from climate change to social instability. That environmental stress could prompt conflicts comes to the fore as the era of environmentalism gathers momentum (such as in, WCED, 1987), with a subsequent expansion of conventional views of security broadened to accommodate such notions (see, e.g., Homer-Dixon, 1991). Essentially, the rationale is that increasing hardship and rivalry over diminishing natural resources and ecosystem services prompts violent conflict, such as potential international 'water wars'. Although this association has common appeal, it did not yield supportable generalisations of the relationship between security and environmental circumstances, despite the availability of exemplars. A report by Schwartz and Randall (2003) to the US Department of Defense describes the risks of international conflict due to the effects of climate change acting in concert with other factors, causing the security implications of climate change to be more widely appreciated.

Opinions are divided over the wisdom of bringing environmental issues into security debates (Trombetta, 2008). 'Securitising' the environment alerts governments to the dangers to the state from environmental destruction, thereby elevating it as 'serious' issue on par with traditional national security concerns in political discourse (Balzacq & Guzzini, 2015). This had several broad effects: *Firstly*, the environment is recognised as a legitimate national security threat, *Secondly*, it offers new incentives for environmental protection and *Thirdly*, it proffers the prospect of considering national security in ways beyond the confines of nation-state territoriality. Sceptics of this development are concerned over the implications of 'militarising' the environmental discourse, with concepts, objectives and practices unsuited to environmental problems. Another aspect of this disquiet is that environmental degradation can be woven into a justification for richer nations to protect their access to resources and economic prosperity. Nonetheless, the notion of *climate security* entered the lexicon of climate change discourse and has remained.

In a major survey, Schubert et al. (2008) conclude that climate change:

- Was not an immediate security risk, but one of medium- to long-term risk that will be realised unless there is successful GHG mitigation,
- Could lead to extreme weather events, food crises, freshwater shortages and greater migration in various world regions that are susceptible to conflicts and crises,
- May cause localised conflicts to spread and that more severe destabilisation could lead to greater conflicts,
- Is unlikely to create classic inter-state wars but is likely to foster areas of destabilisation across borders and existing conflicts will be intensified; it will also generate new conflict threats, and
- Could cause regional destabilisation with global implications, such as by producing unmanageable migration; there are concerns for fragile states becoming overwhelmed and thereby threatening the international system and world economy.

IPCC (2022) offers a less nuanced perspective and states that migration is contributing to violent conflict.

One development is an extension of the security concept from its applications to nation states down to finer scales, even down to individuals, to describe the risks to water and food supply, for instance. It has become commonplace in the climate change discourse to consider food and water supply security as major risks of climate change, and to a lesser extent, energy security (see, IPCC, 2023). Such concerns are now a routine aspect of the human security movement.

Policymakers dealing with risk and with socio-ecological services often recognise individual well-being as that which is at risk from climate change and as a goal when considering risk response measures. Well-being can include avoidance of poverty, good health, material needs being met, opportunities for self-realisation, security and certain social freedoms. Well-being can also be applied at the societal scale, as this is what is usually understood as economic and social development. Although well-being is cast as the benefits to individuals, in large measure these benefits are achieved through public policy, NGO activities and societal institutions directed at society as a whole; in other words, the goal of development is societal well-being, and this is precisely what is at risk from climate change.

CLIMATE RISKS TO NON-MATERIAL SOCIAL/ CULTURAL ENTITIES AND PROCESSES

There has been a tendency in considering climate change impacts on social groups to concentrate on the material aspects of survival and the associated social systems (as reviewed above), overlooking or ignoring the human needs rooted in psychological or psychosocial requirements necessary to maintain social well-being and societal integrity. These intangible benefits encompass a broad array of values (see Table 5.4). Several accounts and overviews of these cultural impacts indicating their associated societal risks are available (see, e.g., IPCC, 2022). Examples of cultural impacts include the loss or damage to fish species/stocks that have symbolic/cultural value; glacial retreat and sense of dislocation; loss of global icons; hunting and fishing declines and erosion of traditional knowledge; degradations of iconic and culturally significant landscapes; diminution of pastoralism as a culture due to changes to seasonality and familiar weather patterns; and the ruin of winter culture and recreation (Adger et al., 2013).

Table 5.4 Types of non-material cultural/societal entities at risk

Societal risk component	Source of benefits	Exemplars of benefits at risk
Non-material cultural/ Societal entities	Direct and indirect intangible outputs from socio-ecological systems	Aesthetic Artistic Heritage Identity Knowledge formation Social cohesion Spiritual Religious

Much of the cultural, emotional and spiritual dimensions of societal well-being are bound to place; climate change can, and is, changing the characteristics of the lifeworld (to use Edmund Husserl's phrase) and is dislocating communities and societies as a result. To a certain extent, while life's necessities can be furnished by any suitable place and that moving a population can enable endangered communities to be removed from evident harms and risks, the cultural, spiritual and other associations between a people and their lifeworld cannot be substituted or replaced without loss. These non-material aspects of cultural/societal value are most associated with traditional societies, but the links between place and the lives of industrial or post-industrial societies cannot be dismissed as being without societal importance.

Describing these intangible aspects of culture can be difficult. By-and-large, the climate change discourse in corporate, governmental, NGO and scholarly realms is dominated by scientific and economic reasoning that has little engagement with immaterial social values. By way of contrast, among communities facing the harms and potential obliteration of these intangible elements of their culture, such concerns are often foremost (see, e.g., Tschakert et al., 2019). Several different arguments are advanced for protecting these values. Societal risks might all be viewed as being highly place- and context-specific, and this applies strongly to these values as well. Societal views determine what of these values is at risk, as different societies may have diverse valuations of the same phenomena. In a sense, these immaterial phenomena underpin every society and are central to a myriad of essential societal attributes. These can be material, such as playing a role in using and managing socio-ecological systems in customs and spiritual practices, education and traditional ecological knowledge. And they can be immaterial, such as by contributing to a collective identity, health and well-being, sense of belonging, social order and cohesion and ways to understand and interpret a lifeworld.

Although these collective social responses are non-material, they are bound up with the materiality of geography and place, and this tie gives rise to not only behaviours, social practices and rituals that are central to specific cultures, but also to material expressions that are labelled as 'cultural heritage' and the like, which are also threatened by climate change. Studies into the risks of climate change to cultural artefacts and heritage materials (i.e., material cultural heritage) are relatively recent, reflecting growing interest (Orr et al., 2021). Such items and places include buildings, heritage sites and objects forming part of cultural practices and societal identity and are important links to the past and foundations for future use. A review by Sesana et al. (2021) used the categories of impacts from: (1) Exposure to the outside environment, (2) Building interiors and their collections and (3) Arising from changes in the physical environment. However, it is considered that this is an emerging field that has yet to extend far beyond an interest in European culture.

CLIMATE RISK TO ECOLOGICAL VALUES

In what might be considered a departure from the conventional framing of risks to social values, the risks to ecological values are included as a component of societal risk. There are two aspects to the case for this inclusion: (1) Ecological values are at significant risks, and (2) That there is a rationale for recognising the risks to ecological values as being a part of societal risks. Regarding the first point, and as canvassed in earlier chapters, climate change is having a devastating effect on ecological values, and this can now only increase in the future with increasing irreversibility, the scope to which depends on the speed and extent of future global warming and associated effects, including sea-level rise and ocean acidification (Adger et al., 2009). For example, permanent ecosystem changes now include glacial retreat and changes to mountain and Arctic ecosystems with permafrost thaw (IPCC, 2023). Biodiversity, a measure of ecological health, is in crisis. Under the IUCN red list, of the 128,918 listed species, 28% face extinction (IUCN, 2020). Within this group, 41% of amphibians, 31% of sharks and 33% of corals have increased extinction risks since 1990 (IUCN, 2020). As climate change continues to affect ecosystems worldwide, biodiversity is lost—the greater the degree of warming, the greater the rate and extent of biodiversity decline.

Both instrumental and intrinsic valuations of ecological systems have been advanced, embracing ecological processes and biodiversity. As biodiversity can embrace genes, species, ecosystems and ecosystem functions, it serves as a useful proxy for considering the risks to ecosystem values as a whole. For a period, there was considerable scholarly attention given to assessing biodiversity in terms of its social benefits, giving rise to an extensive scholarly debate over economic assessments of biodiversity (see, e.g., Ehrenfeld, 1988; Pimentel et al., 1997). Even immaterial goods and services can be assessed for their value of their potential loss to society, such as risks to health and security. In essence, however, such instrumental valuations are identical with those described above for protecting ecosystem goods and services (such as resource harvesting, use of ecosystem services and the like), so that the risks to biodiversity can be viewed as being contiguous with these. Intrinsic valuations of biodiversity offer additional support for ecological values being a part of societal risk, albeit engaging a philosophical rationale, rather than an ecological or economic ones.

Intrinsic valuations recognise that things have value 'for what they are in themselves', regardless of human use or interest. Intrinsic values, therefore, cannot be decreased, granted, increased, substituted or replaced. Exemplars of such values include beauty, dignity, freedom, happiness, life, peace and truth. A number of environmental philosophers propose that biodiversity (and Nature) has intrinsic value (e.g., Rolston, 1988) and the concept is taken up in some international conservation policies. There are differing perspectives on intrinsic values: *Firstly*, that these are conferred socially or personally, based on the values of the assessors, such as given to religious artefacts or totemic species (*subjective intrinsic value*), and *Secondly*, there are valuations completely detached from human attitudes and outlooks and are unconditional, such as we apply to the value of individuals (*objective intrinsic value*), (Sandler, 2012). Support for intrinsic valuations partially arises from the limitations of instrumental valuations of ecology, including that these are:

- Highly selective in attributing value to biodiversity, such as to individual species,
- Highly subjective, as indicated by widely varying valuations of the same phenomena,
- Utilised in decision-making to justify actions resulting in the loss of ecological values, and
- Often criticised as being contingent, replaceable, substitutable and unstable.

Ethical arguments for intrinsic valuations of biodiversity includes that based on social responsibilities. One set of responsibilities occur because society, taken as a whole, and via anthropogenic climate change, is the *source* of the losses of, and ongoing risk to, biodiversity. Furthermore, the other significant drivers of biodiversity losses in the short and medium terms are also of human origin. It follows that society has a responsibility (to the interests of Nature) for this state of affairs, the upshot of which is to recognise that Nature has its own interests and these should be respected as a way to guide social actions towards Nature.

Another argument is that there is no strong case for arguing that humans are the only species entitled to human care. Ethically, there is a justification that society extends its duty of care to some non-human species on a routine basis, most obviously to those species that people 'look after' and for which there are many laws and regulations governing conduct towards such species. One reason that it is difficult to isolate the human species as the sole recipient of the duty of care is that within our own society there are many conditions and circumstances where we acknowledge that individuals require special attention and are not independent ethical agents (such as children and infants). In other words, if it is accepted that all humans are morally equal, then there is no clear moral demarcation from (at least some) other species, as they can experience many of the cognitive and experiential awarenesses and responses of people. It becomes an impossible task to draw a simple line around the limits of human responsibilities where humans are the agents of action: If society has responsibilities towards children and animals, then why not ecosystems? A third stream of argument is that society and Nature are unconditionally entwined when it comes to climate change risks; society and Nature are within the same causal chain. An interpretation of this relationship is that the categories of instrumental and intrinsic dissolve; society needs to recognise and protect natural values as this simultaneously protects social values and vice versa.

Finally, as it is the case the prospects of human societies are inexorably intertwined with that of biodiversity, that the risks from climate change to both are bound by the same logic and cannot be properly separated. As such, the value placed on biodiversity that is at risk must also be in part the value placed on the risks to societies. In such a light, it appears that the instrumental and intrinsic valuations of biodiversity are also as intermingled as they appear to be for societies.

REFERENCES

Adger, W. N., et al. (Eds.). (2009). *Adapting to climate change: Thresholds, values, governance*. Cambridge University Press.

Adger, W. N., et al. (2013). Cultural dimensions of climate change impacts and adaptation. *Nature Climate Change, 3*(2), 112–117. https://doi.org/10.1038/nclimate1666

Arrow, K., et al. (1995). Economic growth, carrying capacity, and the environment. *Ecological Economics*, 15(2), 91–95. https://doi.org/10.1016/0921-8009(95)00059-3

Balzacq, T., & Guzzini, S. (2015). Introduction: 'What kind of theory—if any—is securitization?' *International Relations*, 29(1), 97–102.

Berry, H. L., et al. (2010). Climate change and mental health: A causal pathways framework. *International Journal of Public Health*, 55, 123–132.

Boas, I., et al. (2019). Climate migration myths. *Nature Climate Change*, 9(12), 901–903.

Clement, V., et al. (2021). *Groundswell part 2: Acting on internal climate migration*. The World Bank.

Costanza, R., et al. (1997). The value of the world's ecosystem services and natural capital. *Nature*, 387(6630), 253–260.

Cunsolo, A., & Ellis, N. R. (2018). Ecological grief as a mental health response to climate change-related loss. *Nature Climate Change*, 8(4), 275–281.

Dobson, A., et al. (2006). Habitat loss, trophic collapse, and the decline of ecosystem services. *Ecology*, 87(8), 1915–1924.

Ebi, K. L., et al. (2021). Extreme weather and climate change: Population health and health system implications. *Annual Review of Public Health*, 42(1), 293–315.

Eckstein, D., et al. (2017). *Global climate risk index 2017*. Germanwatch.

Ehrenfeld, D. (1988). Why put a value on biodiversity? In E. O. Wilson (Ed.), *Biodiversity* (pp. 212–216). National Academy Press.

Homer-Dixon, T. F. (1991). On the threshold: Environmental changes as causes of acute conflict. *International Security*, 16(2), 76–116.

IPBES. (2019). *Summary for policymakers of the global assessment report on biodiversity and ecosystem services of the intergovernmental science-policy platform on biodiversity and ecosystem services*. Intergovernmental Science-Policy Platform of Biodiversity and Ecosystem Services (IPBES).

IPCC. (2022). *Climate change 2022: Impacts, adaptation and vulnerability*. Intergovernmental Panel on Climate Change (IPCC).

IPCC. (2023). *Synthesis report of the sixth assessment report (AR6)*. Intergovernmental Panel on Climate Change (IPCC).

IUCN. (2020). *IUCN red list 2017–2020 report*. International Union for Conservation of Nature (IUCN).

Kniveton, D., et al. (2008). *Climate change and migration*. United Nations. https://doi.org/10.18356/6233a4b6-en

Lenton, T. M., et al. (2023). Quantifying the human cost of global warming. *Nature Sustainability*. https://doi.org/10.1038/s41893-023-01132-6

Lin, D., et al. (2018). Ecological footprint accounting for countries: Updates and results of the national footprint accounts, 2012–2018. *Resources*, 7(3), 58.

Maertens, L. (2022). Climatizing the UN Security Council. In S. C. Aykut & L. Maertens (Eds.), *The climatization of global politics* (pp. 143–163). Springer.

McMichael, A. J., et al. (2003). *Climate change and human health: Risks and responses*. World Health Organization.

McMichael, A. J., et al. (2006). Climate change and human health: Present and future risks. *The Lancet, 367*(9513), 859–869.

MEA. (2005). *Ecosystems and human well-being: Synthesis*. Millennium Ecosystem Assessment (MEA), Island Press.

Nordås, R., & Gleditsch, N. P. (2015). Climate change and conflict. In S. Hartard & W. Liebert (Eds.), *Competition and conflicts on resource use* (pp. 21–38). Springer International Publishing. https://doi.org/10.1007/978-3-319-10954-1_3

Orr, S. A., et al. (2021). Climate change and cultural heritage: A systematic literature review (2016–2020). *The Historic Environment: Policy & Practice, 12*(3–4), 434–477.

Pimentel, D., et al. (1997). Economic and environmental benefits of biodiversity. *BioScience, 47*(11), 747–757.

Rolston, H. (1988). *Environmental ethics: Duties and values in the natural world*. Temple University Press.

Sandler, R. (2012). Intrinsic value, ecology, and conservation. *Nature Educational Knowledge, 3*(10), 4.

Schubert, R., et al. (2008). *Climate change as a security risk*. Earthscan.

Schwartz, P., & Randall, D. (2003). *An abrupt climate change scenario and its implications for United States national security*. Defense Technical Information Centre.

Sellers, S., et al. (2019). Climate change, human health, and social stability: Addressing interlinkages. *Environmental Health Perspectives, 127*(04), 045002.

Sesana, E., et al. (2021). Climate change impacts on cultural heritage: A literature review. *Wiley Interdisciplinary Reviews: Climate Change, 12*(4), e710.

Smil, V. (2016). Harvesting the biosphere. *The World Financial Review*, 46–49.

Trombetta, M. J. (2008). Environmental security and climate change: Analysing the discourse. *Cambridge Review of International Affairs, 21*(4), 585–602.

Tschakert, P., et al. (2017). Climate change and loss, as if people mattered: Values, places, and experiences. *Wiley Interdisciplinary Reviews: Climate Change, 8*(5), e476.

Tschakert, P., et al. (2019). One thousand ways to experience loss: A systematic analysis of climate-related intangible harm from around the world. *Global Environmental Change, 55*, 58–72. https://doi.org/10.1016/j.gloenvcha.2018.11.006

UN Environment. (2019). *Global environment outlook—GEO-6: Healthy planet, healthy people*. Cambridge University Press.

Vitousek, P. M., et al. (1986). Human appropriation of the products of photosynthesis. *BioScience, 36*(6), 368–373.

Vos, T., et al. (2020). Global burden of 369 diseases and injuries in 204 countries and territories, 1990–2019: A systematic analysis for the global burden of disease study 2019. *The Lancet, 396*(10258), 1204–1222.

WCED. (1987). *Our common future.* Oxford University Press.

WHO. (2018). *COP24 special report: Health and climate change.* World Health Organization WHO. https://apps.who.int/iris/handle/10665/276405

Xu, C., et al. (2020). Future of the human climate niche. *Proceedings of the National Academy of Sciences, 117*(21), 11350–11355. https://doi.org/10.1073/pnas.1910114117

CHAPTER 6

Climate Change and the Spectre of Collapse

Abstract Societal and civilisational collapse in the past due to climatic events, climate change and related phenomena has captured the attention of the public and policymakers alike, albeit with some controversy. Despite justified scepticism over some historical and pre-historical cases, climate change clearly constitutes a risk to existing and future societies. Tipping points in climate systems pose a particular societal risk, especially for high risk/high probability climate (and related) changes. Climate disasters and extreme events feature prominently in the societal risk of collapse. An account is given of the collapse concept, together with a review of ecological collapse and collapse in the survivalism discourse. A reflection on the collapse discourse, and its significance, closes the chapter.

Keywords Climate change · Collapse · Extreme events · Disasters · Tipping points

There is a long history of concerns over the risks of social collapse, the contemporary version of which is based on failings in vital ecological systems (Haller, 2002), and a number of which feature climate change. Climate has, as Ponting (2007, p. 10) states: "...been a fundamental force in shaping human history", a statement applying equally well to human pre-history. Certainly, climate is a critical environmental factor in the rise and development of civilisations and societies, and sometimes seemingly in their fall. It would be surprising if the advent of the contemporary science of climate change had not spurred speculation as to its effect on the prospects of societies. Societal and ecological collapse has been a part of the climate change discourse from the outset, logically so, given that societal collapse constitutes the ultimate risk from an anthropocentric perspective (Crist, 2007; Richards et al., 2021). As societal collapse represents the most extreme expression of climate risks, its prevention presents as the most compelling argument for minimising future climatic change, for comprehensively predicting its effects and for undertaking the social actions for the necessary adaptations.

Tipping points are central to societal collapse theories and their associated debates. Essentially, a societal collapse is marked by the passing of specific tipping point(s) from which there is no recovery (Richards et al., 2021; Tainter & Crumley, 2011; Wiener, 2018). Societal collapse implies the irreversible, rapid and transformative systemic change that can be brought about by accumulated effects of climate change (and related factors) or by a climatic change brought about by the climate system crossing a tipping point. In other words, societal collapse may be caused by acute or chronic climate change and related impacts. While climate system tipping points are widely recognised, the notion of social tipping points is somewhat controversial and sits unfavourably with some social change theories.

Climate Tipping Points

Although tipping points in the global climate (and related) system have been recognised as an aspect of climate change for some time, their significance has been revised in recent times—and for the worse (IPCC, 2019a, 2019b, 2021). In its early treatments of tipping points, the IPCC considered them only a possibility if extreme future warming took place. Lenton (2011) and others questioned the characterisation of tipping points as high-impact/low probability events, finding that they could be high-impact/high probability events. Relative complacency in mainstream climate science has given way to these concerns with subsequent revisioning. Now the IPCC (2021) reports that such changes could be initiated at the levels of global warming likely to occur even if international efforts to limit GHGs emissions are effective, i.e., a global average warming of 1–2°C or thereabouts. In effect, the possibility of catastrophic climate change is reconfigured from an outside chance towards the distinctly possible; given the conceivable consequences of these events, this amounts to a major reestimation of climate change risks. These reports on scientific research identified the further risk of positive interconnections between these phenomena, such that the occurrence of one event could trigger another, in a causal chain of positive feedbacks.

In many regards, the prospect of such threshold events upsets the risk calculus on which the global climate change agreements rest (Galaz et al., 2016). As to the risks in question, the exact probabilities of the large-scale singularities remain uncertain, albeit being no longer remote. So too are the impacts of such changes uncertain, although 'first principle' deductions suggest that if the impacts are at the high end of the scale, they would be simply catastrophic. This led Lenton et al. (2019) to argue that there is a state of *climate emergency* at the current levels of warming set by the UN FCCC process (i.e., the +2 °C warming target in the 2015 Paris Agreement). In 2020, 11,000 scientists petitioned for climate change, and its impacts, to be considered a global emergency (Lynas, 2020; Rockström, 2020). Many grassroot social movements have echoed this call (Cox, 2020; Rode, 2019a, 2019b, 2019c; Russell, 2019; Thackeray et al., 2020). Significantly, the IPCC concludes that there is a growing share of people recognising climate change as a central risk to society and that view the need for climate action as urgent (2022a).

In terms of risk, the impacts of rapid and extreme climatic changes present the greatest difficulty for prediction, namely forecasting the effects, location, magnitude, and timing of these impacts and therefore, also for planning and conducting a pre-emptive adaptation responses (Galaz et al., 2016). Scenarios of speculative effects of catastrophic tipping points could be devised as a means of determining their associated risks in broad terms, but such inquiries have yet to be conducted to any great extent. Arguably, considering such tipping points involves entertaining the prospect of risks of such ambiguity, uncertainty and potential magnitude, that the logic of conventional risk management approaches would be nullified.

These potential large-scale singularities raise the spectre of existential societal loss (Leemans & Eickhout, 2004; Lenton et al., 2019). Warnings of "…pending disaster are repeated ad nauseam" (Swyngedouw, 2013, p. 9) and eco-catastrophes due to climate change featuring in popular culture may offer an alarmist view of the future, but this threat no longer excites great opposition as a far-fetched possibility as when first produced (Bulfin, 2017; Ripple et al., 2020; Smith, 2022). To the contrary, analogues of potential collapse are regularly presented through extant climate disasters that produce social crises. Contemporary experiences of disasters suggest that passing of tipping points and potential societal collapse is a realistic possibility (Ripple et al., 2020). Rapid and catastrophic environmental change, such as climate change, has led to an increase in the number of recorded natural hazards and in the number of disasters (Leroy, 2006). Disasters and extreme events, therefore, may be interpreted as indicators or precursors of future collapse of ecological and/or social systems or even societies.

Disasters and Extreme Events

Central to the idea of societal risk is the phenomenon of *disasters* and the crises they can precipitate, although disasters and crises are often conflated in the popular imagination and in some research, which can confuse the understanding of climate risks (Boin et al., 2018). Disasters are events (from the realisation of hazards) and crises constitute a social condition that may be a consequence of disasters. In reviewing the disasters discourse, we are concerned with the post-WW2 period after which extensive scholarly inquiry begins (Perry, 2018). Essentially modernist interpretations of disasters focus on societal losses in terms of human life and economic damage.

For most of the history of disaster research and policy responses, disasters are sorted into either those with natural sources (i.e., Of Nature) or resulting from human activity (Perry, 2018; Quarantelli, 2005). In practice, so-called natural disasters (and disaster risks) are clearly amalgams of both categories, as socio-economic and political factors are causal factors in disasters, following the mantra that 'there are no natural disasters, only natural hazards' (see, Wisner et al., 2012; Lopéz-Carresi et al., 2014). Consequently, we have to consider why those societies impacted negatively by a disastrous event "...cannot cope with it" and acknowledge that the reason for this "...lies beyond the natural environment" (Kelman, 2020, p. 16).

Disasters occur when extreme events afflict vulnerable communities. Some communities are relatively unscathed by extreme events because they have the necessary resilience capacities; and vice versa, a similar event can produce disastrous losses in vulnerable communities (i.e., disastrous outcomes), precipitating a crisis (IFRC, 2020; IPCC, 2022a). For this reason, the paired concept of 'disasters and extreme events' arose, as disasters need not be associated with extreme events.

Every disaster and extreme event has unique characteristics. However, it benefits institutions and practitioners in this field to recognise disasters sharing common characteristics for a host of reasons: For facilitating communication, emergency responses, governance, incident management, learning and knowledge exchange, public safety, recovery, risk reduction and other activities (Gundel, 2005; McConnell, 2003). Classifying disasters, however, is difficult due to factors such as the mix of economic, environmental, physical, political, technological and social factors in crises/disasters. As Sementelli (2007) and others have noted, similarities between extreme events are found in circumstances, symbolic representation and response approaches. Classifying disasters is also a function of formal governance and administrative logics (McConnell, 2003).

Disaster preparation, management and response activities (and scholarship) have merged into climate change adaptation preparation and response activities (and scholarship), albeit that the former theme has a far longer history in public policy than the latter (for an overview, see, Glover & Granberg, 2020, ch. 2). In the IPCC (2012) special report on the topic, the risks of climate change altering the characteristics of extreme events is specifically addressed. Climate change will affect the duration, frequency, geographical extent and intensity of events, largely in undesirable ways. Hence, disasters and extreme events have a high profile in the climate change discourse for reasons that are readily appreciated: 1) Actual and potential losses to social and environmental values are large, 2) Economic consequences can be major and, 3) There are great and sometimes permanent changes to societies (Hore et al., 2018). Not all the impacts of disasters are immediate. Disasters and extreme events can create patterns of disruption and weakening of essential social features that can be difficult to discern, but that forebode existential threats and potential societal collapse.

What is 'Collapse'?

Collapsed civilisations, abandoned cities and vanished societies have long enthralled academic researchers and the public at large, with anthropologists, archaeologists and historians prominent in research and scholarship.

Many accounts are replete with inferred or direct lessons on moral failings, social turpitude and the vicissitudes of fate, as famously produced by writers such as Edward Gibbon (1776–1789), Arnold Toynbee (1959) and Oswald Spengler (1962). Anthropologist Joseph Tainter (1988) in *The Collapse of Complex Societies*, made a considerable mark in contemporary scholarship, while Jared Diamond (2005) in *Collapse: How Societies Choose to Fail or Succeed*, was an international best-seller. Despite the seemingly widespread acceptance, if not endorsement, of the societal collapse thesis, it remains controversial within scholarly circles for both empirical and ideological reasons.

In contemporary times, and notably within the ecological collapse discourse, climate change as the dominant causal factor became an increasing theme (e.g., Fagan, 2000; Marshall, 2012; Weiss & Bradley, 2001). Advances in climate change science appear to be partially responsible for this elevation, by revealing the dramatic speed that large-scale future global and regional climate change could occur. This change in scientific and scholarly opinion replaced the previous orthodoxy of gradual climate change (Lenton et al., 2019; van Ginkel et al., 2020). No doubt the rise of environmentalism as a political and cultural movement was also influential, if perhaps more indirectly.

Accounts of civilisational collapse could be said to have Biblical precedents and connections have been made between collapse to both historical and contemporary climate change (Butler, 2016; Tainter & Crumley, 2011; Wiener, 2018), (for a science-based fictional account on the climate induced future collapse of Western civilisation, see, Oreskes & Conway, 2013). While there is a general acceptance that climatic change poses great future societal risks, this awareness is set against considerable differences in scholarly and academic opinion over its role and significance in earlier societal collapses and over whether it might be responsible for future collapse (Wiener, 2014). Further, ideas about collapse are applied to societies, economic systems and ecosystems, but with important differences in meaning and interpretations between these realms (for en overview of potential climate-driven societal collapses, see, Wiener, 2018). In essence, there is disagreement over the responses to the following questions:

- What constitutes a collapse? i.e., What scale of value loss constitutes a collapse? How is a collapse measured or assessed? i.e., How is time considered, such as whether speed of value loss is a defining factor?

- What constitutes a 'society'? i.e., What is the unit or scale of societal entity being considered as being at risk of collapse?
- What causes the collapse? i.e., What causal factors/mechanisms are in play? and
- Does collapse imply a permanent loss?

'Collapse' is one of those somewhat indeterminate terms used in science and policy (Brozović, 2023). It lies between the acceptably loose ideas of a 'disaster', which is essentially a metaphorical description, and more specified and technical terms denoting loss that are commonly understood by specialists in a field. Predictably, and as is often the case with such quasi-technical terms, there are calls within professional groups for greater specificity in its use and for higher-quality definitions, something that, of course, will never happen for a term with such wide common usage. Even among scholars, and perhaps especially so, collapse has varying meanings and interpretations (including ambiguity over what is collapsing); some even doubt its usefulness (see, e.g., Lawler, 2010). Usage of collapse crops up regularly in climate change social science research and policy in reference to economic sectors or social systems. For example, the IPCC (2022a) considers collapse in food systems, health systems, infrastructure systems, livelihoods and of ecosystem services (notably farming and fishing/aquaculture).

Tainter (1988, p. 4) defines collapse in a couple of ways: it is a 'political process' and a "... society has collapsed when it displays a rapid, significant loss of an established level of socio-political complexity". And it must "...be rapid — taking no more than a few decades — and must entail a substantial loss of socio-political structure". This is manifest in many ways, including a lower degree of social stratification and differentiation, less economic and occupational specialisation, less centralised control, less behavioural control, less sharing and trading and less overall coordination. Diamond's (2005, p. 3) perspective is somewhat similar:

> By collapse, I mean a drastic decrease in human population size and/ or political/economic/social/ complexity, over a considerable area, for an extended time. The phenomenon of collapses is thus an extreme form of several milder types of decline, and it becomes arbitrary to decide how drastic the decline of a society must be before it qualifies to be labelled as a collapse.

But there is an important difference in approach here; Tainter emphasises political disruption, whereas Diamond focuses on demographic decline, reflecting an ecological interpretation of collapse. Although archaeological and historical works dealing with collapse reflect both these perspectives, they do not necessarily recognise loss of social complexity or depopulation as markers of societal collapse. Many such works do not define collapse, per se, but authors recognise several characteristics indicative of decline, such as abandonment or diminution of cities or agricultural lands, weakening of economic systems of production and trade, the loss or fragmentation of large political units (such as states or empires) and major weakening of political power and ideological reach. Seemingly, a tendency emerges with writers of 'grand theories' defining collapse in specific ways and offering causal diagnoses, while those advancing more diffuse and multi-sided descriptions of collapse eschew formal definitions of collapse.

It is widely agreed (although not universally) that nearly all the civilisations that ever were, no longer exist—although this prompts the difficult question of how a society relates to a civilisation. But, as suggested above, societal collapse is applied to considerably varying conceptions of society. Middleton (2017) identifies five 'units' of social collapse in this literature: 1) Individual communities, 2) Political units (chiefdoms, dynasties, empires and states), 3) Cultural units, civilisations, ideologies and lifestyles, 4) Systems (including world systems) and 5) Populations and peoples.

What is now taken as a truism—that climatic change can cause civilisational collapse and is therefore the risk of greatest concern—is open to question on several fronts:

- Critics from archaeology and history dispute the climate-collapse discourse on academic grounds (as opposed to being climate change sceptics),
- That the climate-collapse discourse picks up, usually unintentionally, the limits discourse from environmentalism (as neo-Malthusianism), thereby evoking political and ideological issues, and
- That the relationship between social sciences and the natural (biological) sciences raises a number of issues.

Each of these matters is examined below in a review of key climate-collapse issues.

Ecological Collapse

Ecological collapse has become a keystone issue in the sustainability discourse (see the different contributions in, Costanza et al., 2011). Most definitions, or working understandings of ecological collapse, feature a rapid process of ecosystem decline, with major losses of form and structure, and major reductions in their scale and scope with accompanying species losses.

Ecologists and other scientists studying natural biological systems influenced by climatic factors, highlight ecological collapses that have already occurred and the risks, indeed likelihood, of future losses. Nowhere is this more succinctly captured than in the concept of *The Sixth Extinction*, that encompasses losses of animal populations and of keystone species (Kolbert, 2014), with future climate change playing a major role in biodiversity loss (Bellard et al., 2012). Climate change's impact on existing ecosystems and species is documented extensively, such as by the IPCC (see, 2022a), although many large knowledge gaps remain (see, e.g., Pörtner et al., 2021). As this research points out, biodiversity loss has great implications for humanity, given that human life depends on the supply of ecosystem services, so that biodiversity loss diminishes the flourishing of societies and ultimately threatens human survival (as reviewed in Ch. 5 above). Seen thus, extensive and rapid biodiversity loss can be posited as a genuine existential threat to humanity (Chakrabarty, 2009; Tainter & Crumley, 2011; Wiener, 2018).

Therefore, climate change impacts the ability and resilience of natural ecosystems but also threatens society that depends on these systems (Malhi et al., 2020). Furthermore, there are tipping points in many larger ecosystems under stress that, when reached, can cause proportionally rapid collapse (Cooper et al., 2020). There are doubts over whether ecosystem collapse can be predicted, a part of which is the problem of defining and understanding collapse (Sato & Lindenmayer, 2018). As the philosophers Shrader-Frechette and McCoy (1994) point out, sustainability has proved a difficult concept in ecology because two of the key definitional concepts underpinning sustainability—community and stability—have such ambiguous and varying definitions. In effect, these problems carry over that of defining ecological collapse in contemporary times, namely the questions pertaining to the parameters of the system in question (including scale) and to the effects of disturbance of the system. As Boitani et al. (2015, p. 127) argue, while species extinction has a (relatively) unambiguous meaning and is opposed by all, ecosystem collapses are different because they are open systems with "...dynamic composites that change in time and space" so that change can result in varying outcomes with "...little consensus on what is desirable or undesirable".

More specifically, the International Union for Conservation of Nature (IUCN) has a Red List of Ecosystems that identifies those most at risk for biodiversity loss, for which Rodríguez et al. (2015, p. 2) published a guide for its use and defined collapse as.

> ...the endpoint of ecosystem decline, and occurs when all occurrences of an ecosystem have moved outside the natural range of spatial and temporal variability in composition, structure and/or function. This natural range of variation must be explicitly defined in the description of each ecosystem type. Collapse is thus a transformation of identity, a loss of defining features and a replacement by another and essentially different ecosystem type.

Another expression of the threat of ecological collapse occurs under the planetary boundaries concept, tied up with the Anthropocene (see Ch. 1).

Collapse as Survivalism and Societal Collapse

In the history of modern environmentalism, a distinct turning point is the convergence of scientific assessments of the limits of non-renewable natural resource extractions and industrial pollutant acceptance in the natural environment, and the realisation that losing ecological services through industrialisation and mass consumption poses a direct risk not only to human welfare, but to the prospect of the continuance and expansion of the modern project (Guha, 2000). These concerns emerge prominently in the latter 1960s and early 1970s, with classic environmental writings by ecologists and economists, notably *Silent Spring* (Carson, 1962), *The Closing Circle* (Commoner, 1971), *The Population Bomb* (Ehrlich, 1968), *The Limits to Growth* (Meadows et al., 1972), *Small is Beautiful* (Schumacher, 1973), *The Costs of Economic Growth* (Mishan, 1953) and *Toward a Steady-State Economy* (Daly, 1973). Important features of these discourses include:

- Acknowledging there is great inequity in social systems, so that the sustainability project (at essentially any scale of social unit) is inextricably bound into implications and decisions in matters of the distributing the resultant costs, benefits and opportunities for economic development,
- Developing a scientific base for the reckoning of the sustainability concept in which social welfare can be protected through managing resource use, material throughput and waste generation,
- Failing to achieve sustainability, by implication, will result in a failure of modern industrial society, with widespread social and ecological harms, and
- Recognising the social use of natural resources and ecosystem services is subject to the natural laws of ecology (notably from population studies) and thermodynamics (drawing prominently on systems theory).

For a while, the general face of environmentalism was the 'limits to growth' movement and its influence on public awareness, business and industry, education, public policy and law and public administration cannot be underestimated. In essence, the constellation of laws and regulation, institutions, international agreements and other agents of environmental protection and resource conservation, owe their existence to the limits to growth concepts and, to some extent, their associated ideologies (Meadows et al., 1972, 2005). Environmentalism subsequently develops in other directions, prompting other competing discourses, ideologies, political movements and political parties of a wide variety (see, e.g., Carter, 2018; Dobson, 2016). Nonetheless, today's *degrowth* movement is a direct descendant of the limits to growth discourse, although it is more politically explicit, being informed by political ecology and rebuffing accommodations with reformed capitalism, notably rejecting sustainable development (see, e.g., Kallis et al., 2015; Schmelzer et al., 2022).

Particularly relevant to the climate-collapse discourse are the political ideologies and movements arising directly from the limits to growth and degrowth movements, especially that known as *survivalism* (Dryzek, 2005; Katz-Rosene & Szwarc, 2022). There are different strands of survivalism, but all share a central focus on natural ecosystems as being vital for human survival. Re-visiting the key debates over the limits to growth is instructive in understanding the way the climate-collapse controversies have unfolded, and in interpreting their significance.

Donella Meadows et al. (1972) *The Limits to Growth* is a keystone tome; it uses computer modelling (a novel approach at the time) to analyse the relationships between food production, industrial output, pollution, population growth and resource depletion. Interestingly, the consequences of passing the limits to growth are modestly described thus (1972, p. 23): "The most probable result will be a rather sudden and uncontrollable decline in both population and industrial capacity". 'Collapse', however, is used throughout the volume to describe the ultimate outcome of unrestrained growth, with the prospect of a 'dismal, depleted existence'. Industrial capitalism and industrial growth could not continue indefinitely, it is stated, and this tenet encapsulates the zeitgeist of the emergent environmentalism without recourse to the prospective ecological catastrophes that would later come to inhabit the popular imagination. Furthermore, the Meadows et al. (1972) report placed industrial society within an ecological context, such that society cannot experience growth in resource harvesting or population growth in perpetuity (let alone with ongoing exponential growth).

Survivalism built on the proposition that authoritarian powers are a necessity to urgently impose limits on industrial society's appetite for exponential growth (Heilbroner, 1974; Ophuls, 1977), most famously following Garret Hardin's 'tragedy of the commons' thesis (1968)—something that now seems highly anomalous given that green politics subsequent endorsement of democracy, social justice and environmental justice (cf., Biermann & Lövbrand, 2019; Dobson, 2007; Eckersley, 2004). For the most part, the limits discourse was attacked by conservative, right-wing and reactionary political interests with an assorted critique, but one of these themes is particularly relevant to the contemporary climate-collapse debate, namely the charge of being 'neo-Malthusians'.

As Meadows et al. (1972) point out, they are not the first to call for an end to growth, citing their contemporaries Kenneth Boulding, Herman Daly and E. J. Mishan, and also Rev. Thomas Robert Malthus of Georgian England, the author of *An Essay on the Principle of Population* (1798, and later revised). Malthus's essay warned that, in basic terms, unrestrained population growth would outstrip food production growth with resultant famine (and disease and warfare); it was intended as a critique of the Enlightenment's prophets of social progress and of the goal of social equality. Curtailing population growth through 'positive checks' (i.e., chastity and delayed marriage) could bring society into balance with the available supporting resources, argued Malthus; absent of such checks, population size will be reduced (via increased mortality and decreased fertility) to the carrying capacity of available resources.

Industrial society seemingly defied Malthusian pessimism after the Industrial Revolution, thanks to large-scale fossil fuel extraction, material substitutions, productivity gains, technological innovations, the benefits of medical science and sanitation among other factors. In effect, this epidemiological transition is carried forward by modernisation that comprises many elements (Omran, 1971). Problems of scarcity at the broad scale are alleviated by the price mechanism in a globalised economy. Technological advances and colonialisation by imperial states normalised economic growth (and increased net prosperity) in developed nations. Those rejecting the limits to growth thesis and movement founded their views on technological (and economic) optimism.

Dryzek (2005) labels the opponents of the limits thesis as 'Promethians' who, despite not having a coherent ideology or program, are united by a cornucopian belief in the limitless bounty of natural resources and capacity to absorb wastes, albeit that technological advances may be required. Notwithstanding this, business groups, conservative lobbyists and NGOs support promethean causes to promote access to natural resources and resist or prevent environmental protection laws; they are generally opposed state intervention in markets, espousing competitive markets, a belief in self-interest and the primacy of human affairs.

Limits advocates and survivalists evoke the collapse of society when carrying capacity is exceeded, and although much is made of the undesirability of collapse, the concept is often completely undefined. Hardin (1993) *Living Within Limits*, for example, considers a wide array of matters on living within limits but offers little on what collapse means. However, there are writers who are very explicit in describing the natural resources and ecosystem services that gave or could be lost through excessive harvesting and consumption. Brown (1978), for example, describes the threats to the natural resources and biological systems from exponential economic growth, paying particular attention to croplands, grasslands, forests and oceanic fisheries, that he considers having been neglected in the limits discourse and that have become over-exploited and degraded. Using contemporary cases, Brown describes the loss of cropland, deforestation, ocean pollution and overfishing that has destroyed commercial fisheries, the risks of environmental illnesses, climate change risks and endangered species.

Caradonna (2015, p. 15), in a historiography of sustainability, places Diamond (1997, 2005) and similar works (including Tainter, 1988), Morris (2013) and Acemoglu and Robinson (2012) in a historical sustainability category and, although not offering histories of sustainability, do "...reflect deep-seated anxieties about the world's current ecological crisis" and "...are concerned for the fate of modern industrial society". As Cardonna observes, both Diamond and Tainter are interested in the causes of societal collapse, which is the antithesis of sustainability, and their works align with more historical accounts in the humanist tradition. These collapse accounts tend towards the grand theory approach built on empirical exemplars and undertaking comparative analysis.

There are also contemporary political concerns over the climate-collapse discourse. As Swyngedouw (2013), for example, cautions, apocalyptic visions can evoke a 'post-political' condition that universalises societal threats and eliminates debates over the often-extreme social inequities of these threats. Such homogenisation is a danger of aggregation that denies the abilities and agencies of people in responding to climate and the resiliencies that have forged.

What appears to have happened is that there are two versions of the same story being written. One is that derived from ecology and gives rise to contemporary green politics via the debate over the limits to growth. And the other derives from archaeology and history, giving rise to the debate over the collapse of social groups. Neither discourse pays much attention to the other, despite their both being a part of the wider debate over the social risks of climate change.

Diamond's concepts are, in some respects, tailor-made for the era of political environmentalism, as his central thesis concerns the environmental causes of social collapse. His formulation assumes a familiar form for those aware of the limits to growth discourse, wherein human populations reach a size that exceeds the environmental carrying capacity for society (i.e., the *overshoot* condition). Following the Malthusian logic, such societies will (eventually) suffer population collapse and Diamond identifies several case studies. *Collapse*, in contrast to the critical reception within anthropology circles and others, found wide acceptance among those concerned with global environmental issues and has been frequently cited for its descriptions of the societal loss from failing to live within ecological sustainability limits. This latter audience essentially endorse the *limits to growth* thesis underpinning *Collapse*, notably in the light of the threats of climatic change and it is safe to presume are generally unacquainted with the academic debates over the specific case studies of collapsed civilisations that comprise the bulk of the book.

Essentially, the criticism of such views is that they are excessively (environmentally) deterministic and understate the complexity of social change. Many anthropologists reject Diamond's case studies as exemplars, claiming that Diamond was in error and that environmental (and climatic) factors were not responsible. Another more complicating rejection focuses on Diamond's understanding of collapse, by stressing examples of societal transition and of the strengths of resilience in changing societies.

REFLECTING ON THE CLIMATE-COLLAPSE DISCOURSE

Within the discourse of societal collapse, climate change is the major environmental causal factor among a constellation of other environmental and non-environmental factors raised in scholarship (Tainter & Crumley, 2011; Wiener, 2014, 2018). When popular interest in civilisational collapse from climate change arose, it connected with an ancient interest in the fall of civilisations marked by the failure of governments, demise of cultural identity and rise in civic unrest and violence. Civilisational collapse is typically understood as being catastrophic, rapid and involving a loss of support systems, social structure and function.

Although frequently evoked in climate activism and in the mass media, there is, as discussed above, considerable debate and difference within scholarship as to the extent that climatic change caused any historical or pre-historical civilisational collapse—with obvious implications for considering the societal risks of ongoing and future climate change. Despite appearing as a straightforward issue, the risk of civilisational collapse is a convoluted and divisive matter whose importance to societal risk and climate change is a central concern.

Used in the 1990s by some activists and environmental groups, the threat of climate change was depicted as the prospect as the onset of cataclysmic events, a theme taken up widely in books and films in popular culture (Bulfin, 2017). At the same time, it was generally dismissed by governments and scientific bodies. The notion of societal collapse from climatic change has returned as a legitimate concern in contemporary times within anthropology, sociology and other disciplines. It is not immediately clear, however, whether the wide scope of these works reflects disciplinary differences on a complex subject, or whether societal collapse is essentially without any common identity between the disciplines of anthropology, archaeology, history and philosophy and now many additional disciplines (see, Brozović, 2023). Paradoxically, the academic and research literature on sustainability has paid much greater research attention to issues of resilience, than to collapse.

A short review of relevant basic principles might assist in examining the perplexing question of societal collapse. Many of the concepts used in analysing societal collapse are drawn from ecology, including adaptation, carrying capacity, resilience and sustainability. To begin, every social system can be depicted as having a capacity for adaptation when its vulnerabilities to hazards are exposed. As we know that risk is a function of capacity, vulnerability and hazard exposure, climate change adaptations can aim to increase capacity, reduce vulnerability, reduce hazard exposure or all three. After a perturbation or disturbance, the system responses can be seen across a spectrum. In the middle, there is system recovery to its previous state; there can also be an enhanced recovery where performance is improved as a result of the responses taken. Beyond this is systemic transformation, wherein the original system is transformed in a significant way with accompanying improvements in performance. Such outcomes can be viewed as increments of system resilience. At the other end of the spectrum, recovery is only partial and system performance does not return to the pre-shock condition. In the worst of outcomes, the system fails, and no recovery is achieved.

Very few scholars working in this field see climatic change as a singular risk to social viability or explanation for historical and pre-historical social collapses; a social apocalypse occurs when climatic change is concurrent with other contributory factors, which can be social, socio-ecological or natural risks. Yet few of those knowledgeable of the extreme risks facing those societies in the most vulnerable places around the globe would question the likelihood of societal collapse in such locations.

Globalisation and technological change have made the contemporary expressions of the links between societies and socio-ecological systems more complex and interconnected and the extent to which the world's natural systems are exploited for social applications has increased. Accordingly, the vulnerabilities of societies are in some ways greatly increased and in other ways, made more resilient. Increased interconnectivity and the creation of larger-scale social systems poses increased risks that disruptions in subsystems can be multiplied and endanger entire systems. A collapse, therefore, may occur at the end point of several linked factors in a causal chain where the initial perturbation is far removed in time and space from the eventual consequences (and evoking the potential 'cascading, compounding and aggregating' effects). However, the overarching condition of the natural world is that of decline, subject to excessive consumption of natural resources and ecosystem services and degradation of ecological values, resulting in greater vulnerabilities, greater precariousness and lessened resilience of social systems, with concomitant increased risk of societal collapse. In other words, there are clearly extreme hazards to some locations, such that the question is not whether societal collapse is possible in such places, but rather how could a society possibly survive?

REFERENCES

Acemoglu, D., & Robinson, J. (2012). *Why nations fail: The origins of power, prosperity, and poverty*. Crown.

Bellard, C., et al. (2012). Impacts of climate change on the future of biodiversity. *Ecology Letters, 15*(4), 365–377. https://doi.org/10.1111/j.1461-0248.2011.01736.x

Biermann, F., & Lövbrand, E. (Eds.). (2019). *Anthropocene encounters: New directions in green political thinking*. Cambridge University Press.

Boin, A., et al. (2018). The crisis approach. In H. Rodríguez, et al. (Eds.), *Handbook of disaster research* (pp. 23–38). Springer. https://doi.org/10.1007/978-3-319-63254-4_1

Boitani, L., et al. (2015). Challenging the scientific foundations for an IUCN red list of ecosystems. *Conservation Letters, 8*(2), 125–131. https://doi.org/10.1111/conl.12111

Brown, L. R. (1978). *The twenty-ninth day: Accommodating human needs and numbers to the earth's resources*. Norton.

Brozović, D. (2023). Societal collapse: A literature review. *Futures, 145*, 103075. https://doi.org/10.1016/j.futures.2022.103075

Bulfin, A. (2017). Popular culture and the "new human condition": Catastrophe narratives and climate change. *Global and Planetary Change, 156*, 140–146. https://doi.org/10.1016/j.gloplacha.2017.03.002

Butler, C. D. (2016). Planetary overload, limits to growth and health. *Current Environmental Health Reports, 3*(4), 360–369. https://doi.org/10.1007/s40572-016-0110-3

Caradonna, J. L. (2015). The historiography of sustainability: An emergent subfield. *Economic- and Ecohistory, XI*(11), 7118.

Carson, R. (1962). *Silen spring*. Houghton Mifflin Company.

Carter, N. (2018). *The politics of the environment: Ideas, activism, policy*. Cambridge University Press.

Chakrabarty, D. (2009). The climate of history: Four theses. *Critical Inquiry, 35*(Winter 2009), 197–222.

Commoner, B. (1971). *The closing circle: Nature, man, and technology*. Knopf.

Cooper, G. S., et al. (2020). Regime shifts occur disproportionately faster in larger ecosystems. *Nature Communications, 11*(1), 1175. https://doi.org/10.1038/s41467-020-15029-x

Costanza, R., et al. (Eds.). (2011). *Sustainability or collapse: An integrated history and future of people and earth*. MIT Press.

Cox, S. (2020). *The green new deal and beyond: Ending the climate emergency while we still can*. City Lights Books.

Crist, E. (2007). Beyond the climate crisis: A critique of the climate change discourse. *Telos, 4*, 29–55.

Daly, H. J. (1973). *Toward a steady-state economy*. Freeman.

Diamond, J. (1997). *Guns, germs, and steel: The fates of human societies*. Norton.

Diamond, J. (2005). *Collapse: How societies choose to fail or survive*. Viking.

Dobson, A. (2007). *Green political thought*. Routledge.

Dobson, A. (2016). *Environmental politics: A very short introduction*. Oxford University Press.

Dryzek, J. S. (2005). *The politics of the earth: Environmental discourses* (2nd ed.). Oxford University Press.

Eckersley, R. (2004). *The green state: Rethinking democracy and sovereignty*. MIT Press.

Ehrlich, P. R. (1968). *The population bomb*. Sierra Club.

Fagan, B. (2000). *The little ice age: How climate made history, 1300–1850*. Basic Books.

Galaz, V., et al. (2016). Global networks and global change-induced tipping points. *International Environmental Agreements: Politics, Law and Economics, 16*(2), 189–221. https://doi.org/10.1007/s10784-014-9253-6

Gibbon, E. (1776–1789). *The history of the decline and fall of the Roman empire* (Vols. I, II, III, IV, V, VI). Strahan & Cadell.

Glover, L., & Granberg, M. (2020). *The politics of adapting to climate change*. Palgrave Macmillan.

Guha, R. (2000). *Environmentalism: A global history*. Longman.

Gundel, S. (2005). Towards a new typology of crises. *Journal of Contingencies and Crisis Management, 13*(3), 106–115. https://doi.org/10.1111/j.1468-5973.2005.00465.x

Haller, S. F. (2002). *Apocalypse soon?* McGill-Queen's University Press.

Hardin, G. (1968). The tragedy of the commons. *Science, 162*(3859), 1243–1288.

Hardin, G. (1993). *Living within limits: Ecology, economics and population taboos.* Oxford University Press.

Heilbroner, R. L. (1974). *An inquiry into the human prospect.* Norton.

Hore, K., et al. (2018). Climate change and disasters. In H. Rodríguez, et al. (Eds.), *Handbook of disaster research* (pp. 145–159). Springer International Publishing. https://doi.org/10.1007/978-3-319-63254-4_8

IFRC. (2020). *World disaster report 2020: Come heat or high water.* International Federation of Red Cross and Red Crescent Societies. https://media.ifrc.org/ifrc/world-disaster-report-2020

IPCC. (2012). *Managing the risks of extreme events and disasters to advance climate change adaptation.* Cambridge University Press.

IPCC. (2019a). *Global warming of 1.5°c.* Intergovernmental Panel on Climate Change (IPCC).

IPCC. (2019b). *IPCC special report on the ocean and cryosphere in a changing climate.* Intergovernmental Panel on Climate Change (IPCC).

IPCC. (2021). *Climate change 2021: The physical science basis.* Intergovernmental Panel on Climate Change (IPCC).

IPCC. (2022a). *Climate change 2022: Impacts, adaptation and vulnerability.* Intergovernmental Panel on Climate Change (IPCC).

IPCC. (2022b). *Climate change 2022: Mitigation of climate change.* Intergovernmental Panel on Climate Change (IPCC).

Kallis, G., et al. (Eds.). (2015). *Degrowth: A vocabulary for a new era.* Routledge.

Katz-Rosene, R., & Szwarc, J. (2022). Preparing for collapse: The concerning rise of "eco-survivalism." *Capitalism Nature Socialism, 33*(1), 111–130. https://doi.org/10.1080/10455752.2021.1916829

Kelman, I. (2020). *Disaster by choice: How our actions turn natural hazards into catastrophes.* Oxford University Press.

Kolbert, E. (2014). *Sixth extinction: An unnatural history.* Henry Holt and Company.

Lawler, A. (2010). Collapse? What collapse? Societal change revisited. *Science, 330*(6006), 907–909. https://doi.org/10.1126/science.330.6006.907

Leemans, R., & Eickhout, B. (2004). Another reason for concern: Regional and global impacts on ecosystems for different levels of climate change. *Global Environmental Change, 14*(3), 219–228. https://doi.org/10.1016/j.gloenvcha.2004.04.009

Lenton, T. M. (2011). Early warning of climate tipping points. *Nature Climate Change, 1*, 201–209.

Lenton, T. M., et al. (2019). Climate tipping points - too risky to bet against. *Nature, 575*(7784), 592–595. https://doi.org/10.1038/d41586-019-03595-0

Leroy, S. A. G. (2006). From natural hazard to environmental catastrophe: Past and present. *Quaternary International*, *158*(1), 4–12. https://doi.org/10.1016/j.quaint.2006.05.012

Lopéz-Carresi, A., et al. (Eds.). (2014). *Disaster management: International lesson in risk reduction, response and recovery*. Routledge.

Lynas, M. (2020). *Our final warning. Six degrees of climate emergency*. 4th Estate.

Malhi, Y., et al. (2020). Climate change and ecosystems: Threats, opportunities and solutions. *Philosophical Transactions of the Royal Society B: Biological Sciences*, *375*(1794), 20190104. https://doi.org/10.1098/rstb.2019.0104

Malthus, T. R. (1798). *An essay on the principle of population*. J. Johnson.

Marshall, M. (2012). Climate change: The great civilization destroyer. *New Scientist*, *215*, 32–36.

McConnell, A. (2003). Overview: Crisis management, influences, responses and evaluation. *Parliamentary Affairs*, *56*(3), 363–409. https://doi.org/10.1093/parlij/gsg096

Meadows, D., et al. (2005). *Limits to growth: The 30-year update*. Chelsea Green Publishing.

Meadows, D. H., et al. (1972). *The limits to growth*. Universe Books.

Middleton, G. D. (2017). *Understanding collapse: Ancient history and modern myths*. Cambridge University Press.

Mishan, E. J. (1953). *The costs of economic growth*. Staples Press.

Morris, I. (2013). *The measure of civilization: How social development decides the fate of nations*. Princeton University Press.

Omran, A. R. (1971). The epidemiologic transition: A theory of the epidemiology of population change. *The Milbank Memorial Fund Quarterly*, *49*(4), 509–538. https://doi.org/10.2307/3349375

Ophuls, W. (1977). *Ecology and the politics of scarcity*. Freeman.

Oreskes, N., & Conway, E. M. (2013). The collapse of western civilization: A view from the future. *Daedalus*, *142*(1), 40–58. https://doi.org/10.1162/DAED_a_00184

Perry, R. W. (2018). Defining disaster: An evolving concept. In H. Rodríguez, et al. (Eds.), *Handbook of disaster research* (pp. 3–22). Springer. https://doi.org/ https://doi.org/10.1007/978-3-319-63254-4_1

Ponting, C. (2007). *A new green history of the world: The environment and the collapse of great civilizations* (2 ed.). Vintage Books.

Pörtner, H.-O., et al. (2021). *Scientific outcome of the IPBES-IPCC co-sponsored workshop on biodiversity and climate change*. IPBES secretariat.

Quarantelli, E. L. (Ed.). (2005). *What is a disaster?* Xlibris Publishers.

Richards, C. E., et al. (2021). Re-framing the threat of global warming: An empirical causal loop diagram of climate change, food insecurity and societal collapse. *Climatic Change, 164*(3), 49. https://doi.org/10.1007/s10584-021-02957-w

Ripple, W. J., et al. (2020). World scientists' warning of a climate emergency. *BioScience, 70*(1), 8–12.

Rockström, J. (2020). Why we need to declare a global climate emergency now. *The Financial Times* (July 28, 2020). https://www.ft.com/content/b4a112dd-cafd-4522-bf79-9e25704577ab

Rode, P. (2019). *Climate emergency and cities: An urban-led mobilisation? The climate decade's priorities for urban climate action, policy and research*. London School of Economics.

Rodríguez, J. P., et al. (2015). A practical guide to the application of the IUCN red list of ecosystems criteria. *Philosphical Transactions B, 370*, 20140003. https://doi.org/10.1098/rstb.2014.0003

Russell, B. (2019). Beyond the local trap: New municipalism and the rise of the fearless cities. *Antipode, 51*(3), 989–1010. https://doi.org/10.1111/anti.12520

Sato, C. F., & Lindenmayer, D. B. (2018). Meeting the global ecosystem collapse challenge. *Conservation Letters, 11*(1), e12348. https://doi.org/10.1111/conl.12348

Schmelzer, M., et al. (2022). *The future is degrowth: A guide to a world beyond capitalism*. Verso.

Schumacher, E. F. (1973). *Small is beautiful: A study of economics as if people mattered*. Blond & Briggs.

Sementelli, A. (2007). Toward a taxonomy of disaster and crisis theories. *Administrative Theory & Praxis, 29*(4), 497–512. https://doi.org/10.1080/10841806.2007.11029615

Shrader-Frechette, K. S., & McCoy, E. D. (1994). How the tail wags the dog: How value judgements determine ecological science. *Environmental Values, 3*(2), 107–120.

Smith, C. (2022). Climate change and culture: Apocalypse and catharsis. *Ethics & the Environment, 27*(2), 1–27.

Spengler, O. (1962). *The deciline of the west*. Knopf.

Swyngedouw, E. (2013). Apocalypse now! Fear and doomsday pleasures. *Capitalism Nature Socialism, 24*(1), 9–18. https://doi.org/10.1080/10455752.2012.759252

Tainter, J. A. (1988). *The collapse of complex societies*. Cambridge University Press.

Tainter, J. A., & Crumley, C. L., et al. (2011). Climate, complexity, and probelm solving in the roman empire. In R. Costanza (Ed.), *Sustainability or collapse: An integrated history and future of people and earth* (pp. 61–75). MIT Press & Dahlem Workshop Reports.

Thackeray, S. J., et al. (2020). Civil disobedience movements such as school strike for the climate are raising public awareness of the climate change emergency. *Global Change Biology*, 1–3.

Toynbee, A. J. (1959). *A study of history*. Oxford University Press.

van Ginkel, K. C. H., et al. (2020). Climate change induced socio-economic tipping points: Review and stakeholder consultation for policy relevant research. *Environmental Research Letters, 15*(2), 023001. https://doi.org/10.1088/1748-9326/ab6395

Weiss, H., & Bradley, R. S. (2001). What drives societal collapse? *Science, 291*(5504), 609–610. https://doi.org/10.1126/science.1058775

Wiener, M. H. (2014). The interaction of climate change and agency in the collapse of civilizations ca. 2300–2000 bc. *Radiocarbon, 56*(4), S1–S16. https://doi.org/10.2458/azu_rc.56.18325

Wiener, M. H. (2018). *The collapse of civilizations*. Belfer Center for Science and International Affairs.

Wisner, B., et al. (Eds.). (2012). *The Routledge handbook of hazards and disaster risk reduction*. Routledge.

CHAPTER 7

Climate Change, Politics and the Transformation Challenge

Abstract Politics is a key factor in determining the societal construction of risk, its use and the consequences of that use. Societal transformation is central in responding to the political challenges of societal risk but requires understanding these political factors. Three forms of political inquiry are used to explore these risks: Political ecology, social justice and climate justice. Political ecology directs attention to the environmental politics related to such matters as access, ownership and use of resources and services and their implications for society and the environment. Social justice inquiries explicitly deal with the distribution outcomes, procedural justice and recognitional justice arising from climate change risks. Climate justice takes these concerns further, raising the ethical and moral dimensions of the differentiation of climate risks within and between societies.

Keywords Climate change · Climate justice · Equity · Politics · Political ecology · Social justice

Politics shape societal risks from climate change. Pre-existing social conditions determine the societal outcomes of any given climatic and related hazard for any given location, together with the actions taken post-event. Social and economic differentiation predisposes the distribution and allocation of societal climate risk, producing different risk vulnerabilities and risk exposures within and between social groups and locations. Understanding the political factors that create, maintain and perpetuate socio-economic distributions of education, health, opportunities, power and influence, mobility, security, social status, wealth and other factors provides insights into the differentiation of risk. By implication, significantly reducing the climate (and related) dangers facing the most vulnerable societies may involve scientific and technical interventions but it invariably evokes the need for political interventions and actions. Transformations as a risk reduction strategy are a form of *proactive or anticipatory adaptation* to climate change and social transformation has been applied most frequently in the adaptation context (see, e.g., Pelling, 2011).

Knowing who is at risk (and why) is inescapably a political inquiry educing the concerns of equity and justice. This also entails an interest in societal transformation to mitigate vulnerabilities created by structural socio-economic injustices. In order to mitigate vulnerabilities embedded in societal structures "...the persistent effects of unjust advantage from the past on social and economic relations" must be taken into account (Meister, 2002, p. 97).

But what is meant by social 'transformation'? In broad terms, transformation involves changes to the essential character of a system, which can mean altering such elements as its forms, functions, meanings, relationships and structures. Transformation in the climate change context is therefore multifaceted and complex, spanning governmental institutions, jurisdictional realms, policy domains, socio-ecological systems and social divisions at a range of scales. When applied to social systems, transformation results from intentional choices or as impositions by changing external circumstances. As such, transformation is subjective with its outcomes assessed according to chosen value sets.

Transformation, by definition, expresses social change and therefore elicits many political interests and issues (and this certainly the case for the climate change discourse). There are competing views as to what the politics of transformation comprise and whether it concerns political processes, political outcomes or reforms to political institutions and systems (Manuel-Navarrete & Pelling, 2015). Many place transformation at the centre of the climate change risk discourse (e.g., Pelling 2011).

Politics aiming at enabling transformation in ways that can mitigate societal climate risk and vulnerability can be understood as a 'normative project'. A growing scholarship argues that transformative forms of governance are essential to catalyse the necessary social change to secure a progressive social agenda for climate change responses (Bosomworth et al., 2017; Jensen et al., 2020; Rickards, 2013). Substantially transforming the cultures, norms, processes, systems and values underpinning and reinforcing currently unsustainable practices and development pathways is necessary for achieving *climate just* futures (Moloney et al., 2018). This means that transformations, in essence, are political activities involving political actors, institutions and values and producing political outcomes. Because of these and other factors, politics has proved a challenging matter for transformation scholarship which tends to adopt a narrow interpretation of its influence and role; in many instances, politics is viewed in institutional forms and where it is deemed to stifle transformation endeavours.

There are many contributions in the environmental and climate politics literature addressing these issues. This scholarship takes many forms with several typologies available for understanding the connections between risk, vulnerability, values and transformation. Here, we are particularly interested in *political ecology, social justice* and *climate justice*. Not only are these forms prominent in climate change discourse, but each makes a distinctive contribution to the risk and transformation discourse.

POLITICAL ECOLOGY

In some respects, political ecology is a fluid category with interests in the inter-relationships between ecology, the environment and politics, and can include environmental justice and political economy of the environment, with some scholars in these fields using the terms interchangeably (Robbins, 2012; Sovacool & Linnér, 2016; Taylor, 2015). As such, it tends to be recognised by a core set of concerns and interests, with its definitional boundaries or limits being widely interpreted which are, accordingly, uncertain. As with climate justice (discussed below), political ecology thematically describes types of environmental activism and variants of environmental discourse, scholarship and theory. Political ecology has also been adopted by inquiries and studies far removed from environmental issues, such as in criminology, education, health and labour studies, and sometimes as an approach to investigate the influence of specific circumstances/locations as the locus for particularised social relations. As a term and concept, political ecology emerged in several places and across different fields of inquiry during the 1970s. Blaikie and Brookfield (1987) explicitly place ecology and political economy into the same frame of reference and articulated the dynamic relationship between societies and their ecological setting and that also within and between groups comprising society.

Vaccaro et al. (2013), along with others, describe the disciplinary foundations of political ecology as cultural studies of the environment, political anthropology and geography, and political economy, with each bringing forth particular concerns (see Table 7.1). Such multi-disciplinarity has broadened further and now political ecology embraces a wide array of disciplines and interests; this has been accompanied by scholarly debate over the identity of political ecology.

Hence, political ecology has come to assume a diverse character and critically, it is without a single and central theory. Political ecology, however, has a keen interest in politics regarding "...environmental access, management, and transformation" (Robbins, 2012, p. 3) and was part of the critical response to the scientific-technical orientation of IPCC's understanding of vulnerability (Malone & Engle, 2011; O'Brien et al., 2007). Watts (2000) describes three approaches undergirding the beginnings of political ecology, namely: (1) Systems theory applied to ecology/cybernetics, (2) Ecological anthropology/cultural ecology and (3) Natural hazards/disaster research. Paulson et al. (2003), describe a set of core concepts: (1) Marginality: In which ecological, economic and political factors combine to produce conditions of social marginalisation, (2) Unsustainable resource use: That occurs as a consequence of social

Table 7.1 Political ecology: Disciplinary contributions

Discipline	Contribution
Cultural studies of the environment	Cultural heritage and landscapes, cultural transformations of Nature, identity politics, taste
Political anthropology and geography	Governmentality, sources of legitimacy, state making, territoriality
Political economy	Commoditisation, gentrification, market integration, niche markets

Source Vaccaro et al. (2013)

relations and (3) Social relationships with the environment: That vary according to interests, perceptions, positions and rationalities.

Accordingly, the core of political ecology concerns social relations related to access, ownership and use of resources in analysing environmental problems and in articulating the case for conservation and environmental sustainability within a context of social justice (Kasperson & Kasperson, 2005; Sze, 2021). Particular attention is given to the subject themes of class, culture, environmental degradation, markets, socio-economic disadvantage, poverty, traditional knowledge, resource access and use, states, vulnerability and wealth. As Roberts (2020) observes, political ecology identifies the role of international development and modernisation programs on local communities and environments in developing nations (such as logging, mining and Nature conservation programs), highlighting the role of markets and states in the acts of damage, disturbance and dispossession.

Critiques of political ecology have played a role in its continuing evolution. As Watts (2000) explains, these concerns promote greater attention to the questions of governance, politics and power through examining questions of community and governance, entitlements, gender and resistance. Even a cursory examination of research publications self-identifying with political ecology bears out this range of critiques; many papers are entirely devoid of references to ecology other than the subject matter and similarly, there are those wanting of any direct reference to politics (see the discussion in, Glover & Granberg, 2020, pp. 144–146).

Political Ecology and Climate Risk

Political ecology engages with climate change risk across a range of issues and subjects, reflecting the diversity of the concept and wide applicability across the climate change discourse (cf., Goldman et al., 2018; Sovacool, 2021; Taylor, 2015). In many respects, political ecology is the source of much of the progressive critique and inquiry into climate change risk, especially where these evoke political considerations.

There are numerous examples of such studies. Taylor (2015) argues that climate change should not be considered as an external threat to agrarian livelihoods, but one of the risks generated by global capitalism and the inequitable distribution of socio-ecological resources. Eriksen et al. (2015) similarly make a case for re-framing climate adaptation in order to recognise the political construction of climate risks so as to reveal the 'socio-political causes' of vulnerability. Sovacool and Linnér (2016) and Sovacool (2018) identify four ways in climate adaptation, using a range of case studies, in which socio-economic inequality within social groups is increased, namely *enclosure* of resources and means by vested interests (such as by privatisation), *exclusion* of social groups from decisions involving their interests, *encroachment* that reduces environmental values and *entrenchment*, whereby women and minority groups are disempowered. Agarwal et al. (2012) analyse climate adaptation projects across several developing nations and identify lessons for decentralising resource governance, including increasing local capacities through transfers of information, finance and technical resources and through community empowerment in planning and implementation. Demetriades and Esplen (2009) describe the intersection of gender and poverty in the construction of climate change risks and the particular problems faced by women; similarly, Nightingale (2011) finds that gender intersects with other factors, such as race, in producing such risks and injustice.

Political ecology views climate change as one central result of humanity's "...restless compulsion to transform nature" (Taylor, 2015, p. 1). At the same time, social structures have been transformed in ways that increase inequity and create an asymmetric distribution of climate change vulnerability in societies. Hence, a transformation of these social structures and related value systems is necessary to mitigate the risks from climate change and to allocate costs and benefits in more equal ways (cf., Pelling et al., 2015).

SOCIAL JUSTICE

Social justice concerns allocating and distributing benefits and disbenefits within society according to normative principles, adjudging what is fair or just (Fünfgeld, 2019; Levy & Patz, 2015; Kaswan, 2021; Malloy & Ashcraft, 2020; Marion Suiseeya, 2021). Ethical, practical and legal reasons are employed to support social justice causes. Some recognise three elements in social justice: (1) Distributive justice, (2) Procedural justice and (3) Recognitional justice (as interactional or interpersonal justice). Here, we consider recognitional justice as part of procedural justice. In the seminal *A Theory of Justice*, John Rawls (1971) argues that social justice requires distributional and procedural justice. Specific conceptualisations of society are central to social justice and Miller (1999) identifies three necessary assumptions for theorising about social justice:

(1) A society that is bounded or delimited, (2) Social institutions capable of facilitating social justice initiatives for individuals and (3) An agency capable of influencing these institutions according to the demands of social justice theory.

In contemporary societies, the state is the primary institution of social justice in its distributive and procedural forms (Barry, 2005; Harvey, 2009). Much of the positive and normative argumentation over social justice entails the composition of social institutions and the state, namely their aspirations, identity and norms. Political analysis and political values feature strongly, therefore, in considerations of social justice. Benefits and positive opportunities and privileges can assume material and service forms, typically identified as being associated with employment, income, leisure and recreation, property, resource access, security and provided services, such as elderly care, childcare, culture and entertainment, education, health and mobility. Social justice can also consider that which influences welfare, including access and use of political power and influence and the effects of political oppression and discrimination. Benefits are valued because they provide opportunities to enhance individual and collective contentment and welfare, although social justice is preoccupied with providing the means of achieving higher welfare, not its actual fulfilment, which is generally considered as being the responsibility of the recipients.

Answering Harold Lasswell's (1936) classic question of 'Who gets what, when and how?' (this is also one of the most-cited definitions of politics and also a central argument for the study of politics) with the additional question of 'Why?' (Kaswan, 2021), is at the centre of the *distributional* aspect of social justice, if not at the centre of the identity of the social condition. Many factors and accompanying rationales are used in explaining the distribution of benefits/disbenefits, including abilities, capacity to utilise, economic efficiency, needs, responsibilities in society and rights. Equitable allocations can be considered across society (*horizontal equity*) so that benefits are equivalent to socio-economic position (i.e., treating people in the same circumstances equally) and considered through society (*vertical equity*), in which resources are allocated according to need (i.e., those with greater responsibilities receive greater rewards). Other schemes are equal shares for all and the satisfaction of basic needs. Failures in social justice are marked societies subject to discrimination, exploitation, oppression and tyranny.

Procedural justice has attracted greater attention in recent times, featuring calls for greater participation in climate decision-making from groups previously excluded, including indigenous peoples, minority populations and women (cf., Klepp & Chavez-Rodriguez, 2020). Ethically, the cases made for procedural justice have features in common with those advanced for distributional equity and fairness. Stakeholders, under a range of ethical schemes, are recognised as having the right to participate in decisions in which they have an interest or a share in a collective interest, thereby promoting self-actualisation (such as expressed in individual liberty). Awareness and understanding of others' needs and wishes is also of moral worth and this builds social bonds and establishes the foundations for communication and the search for solutions of mutual benefit. Securing such ideals in social conditions as they exist is difficult, as shown by the differences between the theories and practices of democracy (see, e.g., Parvin & Saunders, 2018).

There is a voluminous literature on this topic, but a summary will suffice here (see Table 7.2) (Begg, 2018; Glicken, 2000; Hügel & Davies, 2020; Reed, 2008). Within the concept of stakeholder inclusion is a wide range of relationships from the passive to the highly involved, covering consultation, involvement and participation, with the key differentiating aspect being the extent to which power sharing occurs in the decision-making process. Participation extends across the decision-making process to include not only contributions to knowledge accumulation and knowledge applications, but also to designing and conducting the decision-making process itself. Further, there is a myriad of institutional forms involved, including those providing advising, co-ordinating, funding, implementing and regulating functions.

Social Justice and Climate Risk

Arguably, social justice is not a conventional perspective, theme or type of climate change risk discourse and is not one that is easily confined within clear conceptual boundaries or derived from a singular scholastic lineage (such as cultural risk theory or Normal Accident Theory). On the other hand, one concern emanating across effectively every aspect of climate change risk discourse is that of justice, particularly social justice (see Jafry, 2019). To review the range of perspectives on climate change risk and omit social justice would be to leave out the animating aspect of many differences and debates over the societal risks of climate change.

Table 7.2 Pragmatic benefits of stakeholder inclusion in risk decision-making

Benefit	Risk implication
Input of local/indigenous/traditional knowledge overlooked by orthodox scientific-technical approaches	Expands risk knowledge field; may influence outcomes
Articulates the different perspectives of stakeholders	Expands risk knowledge field; may influence outcomes
Ensures issues of stakeholder importance are included	Includes risks perceived by stakeholders
Ensures that all stakeholder groups have an opportunity to be involved and to express their views (as opposed to having these overlooked or expressed by others)	Stakeholders can directly articulate their risk issues, priorities and views, and of decision-making processes
Provides opportunities for authorities to respond to stakeholders concerns and interests	Authorities have to opportunity to respond to other stakeholder's inputs on risk
Aids communication and information exchange between authorities and other stakeholders	Knowledge exchange on risk between all stakeholders
May identify assessment and management options previously unknown/overlooked by authorities	Expands risk knowledge field; may influence outcomes
May help in the ranking/evaluation/ measuring of risks	Expands risk knowledge field; may influence outcomes
Can help build credibility/legitimacy of all knowledge forms by participants	May address any risk knowledge ignorance, misconceptions and falsehoods held by any/all stakeholders
Can assist in building consensus in decisions reached	May assist in risk management, other risk responses and risk communication

Including social justice in this review does, however, pose some practical challenges in view of the diversity and complexity of the issues involved. A simple schematic arrangement of social justice issues recognises the influence of socio-economic differences in *risk exposure* within society (i.e., the social pre-determinates of vulnerability to climate change impacts), the influence of these in *representation* in decision-making processes in response to risks (such as participation in public policy and planning processes) and the influence on the *consequences* of adaptation responses (see, e.g., Kaswan, 2021). Risk reflects prevailing circumstances, the character of hazards and the consequences of hazard occurrence, so that it indelibly expresses social justice. Many of these concerns mirror those raised in disaster risk management and in sustainable development policies and practices, and these discourses have influenced the climate change risk discourse and vice versa.

Much of the scholarship on climate change adaptation deals with social justice themes, particularly the differential effects of climate change impacts, development strategies and the effects of natural disasters on social groups (Kasperson & Kasperson, 2005; Levy & Patz, 2015). Nearly all this work adopts a positivist outlook, meaning that it analyses and records the iniquitous outcomes of these circumstances and events, especially for cultural minorities, indigenous and traditional societies and the socio-economically disadvantaged. Such groups are (usually) the most vulnerable to natural and technological hazards. Pursuing social justice, or avoiding increasing social injustice, provides the rationale for these inquiries. Only in a minority of cases, however, is this research and scholarship explicit in its normative program. In other words, expressions of the aspirations for social justice are absent and accordingly, its identity is inferred and understood indirectly by describing what are deemed explicitly or implicitly as *social injustices* (i.e., akin to *negative evidence* in science).

Stakeholder inclusion, as discussed above, promotes an expression of the democratic ideal and one enshrined in numerous legislative and regulatory regimes (although in many respects, this rationale is a late arrival in supporting participatory practices in risk-related decision-making processes). Stakeholders can be an important source of local knowledge drawing on indigenous/traditional knowledge of natural systems, social/cultural adaptations and other responses to these systems' hazards (ILO, 2019; IPCC, 2014). Risk responses at the local level are often dynamic, so that understanding the triggers and thresholds in risk management activities offers insights into the cultural dimensions of risk knowledge and of the social implications of hazards (such as ecosystem service reliability, health risks and resource availability and supply).

Climate change has been recognised as threatening social welfare since the inception of the climate change discourse (Gough, 2010; Johansson et al., 2016), thereby evoking the protective responses of international bodies, NGOs and states, such as expounded by the UN FCCC. It is difficult to capture the broad range, complexity and depth of the potential social harms and the extent to which these would further social inequity and lower the welfare of those at the lower tiers of the socio-economic spectrum in simple terms, although a sample of relevant scholarship provides some indications.

Global inequalities between nations are exacerbating under climate change and forecast to continue and increase (see, e.g., IPCC, 2023; Mendelsohn et al., 2006; Tol et al., 2004). Diffenbaugh and Burke (2019), for example, find that global warming increases global economic inequality, partly through economic declines experienced by hotter and poorer nations. Islam and Winkel (2017) identify three primary sources of increased inequality from climate change impacts: (1) Increased exposure of disadvantaged groups to adverse climate effects, (2) Increased susceptibility of such groups to losses and (3) Reduced coping and recovery from such impacts. Their work emphasises that disadvantaged groups will suffer proportionally greater losses than the social average and that the three elements above are linked in a vicious cycle, whereby the losses from climate hazards increase vulnerabilities to future climate. Although the effects of disasters often attract the greatest attention, the impacts of shifts in prevailing climatic conditions over time can have significant effects on ecosystem services and those whose livelihoods and welfare are closely tied to primary production, such as changes in rainfall patterns and water supply, the timing of seasonal rains, temperature extremes and the like (see, e.g., IPCC, 2023).

Asset damage and losses (including the financial, human, physical and social types) and income declines can combine with reduced access to biophysical resources and ecosystem services to cause catastrophic social and economic instability. Lower income within a social group or formal jurisdiction is tied to higher vulnerability to climate change impacts. Increased social and economic inequality is also a marker for overall poorer levels of environmental sustainability. Arguably, social and economic inequality in a social group makes it more vulnerable than an equivalent group with greater socio-economic equality. In such circumstances, the poor have fewer resources and societies exhibiting higher inequality in socioeconomic measurements provide fewer public resources for the poor for reducing their risk exposure, vulnerabilities and other risk management responses. Again, this highlights the need for transformative change in more equitable directions in order to mitigate risks related to climate change (Patterson et al., 2018).

CLIMATE JUSTICE

While many of the prominent risk theories presented earlier in this volume have their origins in scholastic writings and academic reflections on social conditions, environmental and climate justice (hereafter known as *climate justice*) presents ideas largely emanating from the activities of social movements and environmental activism by environmental NGOs and community organisations (see, e.g., Jafry, 2019; Skillington, 2019; Walker, 2012). By way of navigating through contested definitions, it is helpful to consider environmental inequity as an overarching social condition, characterised by specific problems, such as environmental racism, the resolution of which constitutes climate justice (see the different contributions in, Jafry, 2019). Climate justice shows that climate change is not just a scientific-technical or financial issue, but entails ethical and moral judgements about justice; climate justice "...helps to reframe mainstream debates to usher in critical attention to social impacts, outcomes, and justice concerns" (Sultana, 2022, p. 118). Furthermore, climate justice creates a framework for "...an engaged grassroots response to the unfolding climate crisis" (Tokar, 2019, p. 13).

Social practices continue to inform the ideas and concepts of climate justice, giving it a dynamic and evolving character, another feature that distinguishes it from the more academic perspectives on risk that tend to be comparatively static (Carr & Nalau, 2023). Politically and sociologically, climate justice has another distinguishing feature—that of its origins in movements of minority populations and low-income groups, contrasting with the domination of the middle-class and professional membership in environmental NGOs in developed nations responsible for environmental activism and political campaigns. Equally significant is that women hold leadership roles in this activism in both developed and developing countries (see, e.g., Krauss, 1994, 1998; Sultana, 2022; Terry, 2009).

A central aspect of climate justice is its intergenerational dimension, wherein contemporary decisions need to include the interests of future generations, i.e., their legacy implications (Howarth, 2013; Meyer, 2016; Page, 1999). Contemporary action, or inaction, not only affects contemporary individuals and society but also those not yet born. Climate change is clearly a threat to the long-term survivability of both social and ecological systems and, accordingly, poses severe risks for future generations (cf., IPCC, 2023; Skillington, 2019). Protecting climate benefits for future generations is identified in the first Principle of the UN FCCC. This is, in many ways, also the core of sustainable development as formulated in the famous definition by the United Nations' WCED (1987).

Most of the writings on environmental justice concern the US and several histories record the rise of the concept there (Murdock, 2021). Protests over environmental hazards by local communities of racial minorities and those of lower socio-economic status occurred through the 1970s and 1980s in the US. Most of these hazards were created through industrialisation, including contaminating drinking water, releasing toxic pollutants to the environment and using agricultural biocides harmful to farm workers. In 1984, the US General Accounting Office published a report that examines the links between hazardous-waste landfills and the racial and socio-economic profiles of nearby communities, establishing a positive association with racial minorities and of low incomes (GAO, 1984). Another landmark study systemically links the location of public waste sites to places dominated by minority populations across the US (CRJ, 1987). Critically, race was a more influential variable than any other in explaining the location of waste dumps, including that of low socio-economic status (Bullard, 1990).

Climate Justice and Risk

This dialogue opened a new discourse on risk, particularly its allocation and distribution (Kaswan, 2021). Happenstance does not account for the unequal social distribution of risk, but rather it is the consequence of social practices. In effect, the environment is an agent of risk in a social program of prejudice (including against age, class, culture, ethnicity, gender, health status, race and religious beliefs) through inequitable, unfair and unjust social practices. Routine social life can be characterised by the benefits of a relative freedom from certain risks and dangers, such as arising from consuming imported goods and services requiring hazardous production, the location of settlements and from working environments. Critically, the social ideal may also be to enjoy a low vulnerability from climatic hazards.

Many societies and many millions of citizens face a prospective future far from this social ideal. Climate change will deepen existing social inequities and inequalities within and between societies and expose societies to new risks. At the extreme end of these impacts is societal collapse for societies already perilously exposed to climatic and related risks.

When climate justice has been codified in regulatory terms by government agencies, there is typically a reference to ensuring uniformity in society in the protection provided from climate risks and hazards, sometimes explicitly recognising the need to protect minority groups and those of low income. Such deliberate measures may also include notions of procedural justice, as well those of distributive justice. What climate justice has brought to the broader risk discourse is the theme of responsibility and accountability. If risks are imposed differentially within society, then this is the result of decisions by corporations, governments and institutions. This means that existing "...structures, institutions, habits and priorities need to be critically re-evaluated in light of the risks that climate change poses" (Rickards, 2013, p. 690). Hence, any real climate justice needs a transformation towards an equitable distribution of climate risk.

POLITICS, JUSTICE AND THE POTENTIAL FOR SOCIETAL TRANSFORMATION

As discussed elsewhere in this volume, climate change is in itself a global, disrupting and transformational agent with broad implications for society, managed ecosystems and the natural world. As the three perspectives discussed above highlight, achieving just climate risk futures requires transformations in institutional cultures, norms, systems and values. As such, this requires acknowledging those societal factors producing climate injustice resulting in uneven capacities and vulnerabilities, such as asymmetric cost and benefit allocations, power asymmetries and social stratification. As the literature on political ecology, social justice and climate justice shows, transformative interventions are needed to catalyse such shifts through changes in social contracts for progressive social goals; this involves addressing the social root causes of vulnerability. Transformative change following climate disasters and extreme events can be indicative of the failure of the extant social contract involving the protection of individuals and groups in society (Adger et al., 2013, 2019; Hayward & O'Brien, 2010). Applying the social contract can be problematic, however, especially given the difficulty of defining rights and responsibilities in real world conditions with diverse actors, power asymmetries and multiple scales of governance.

Pursuing climate justice for societal risk exposure, necessitates recognising that, as a complex adaptive system, public policy agendas are shaped through dominant beliefs, frameworks and ideologies (Bosomworth,

2018). Radical transformation requires challenging existing cultures, dominant rationalities and habitual practices that are institutionalised within governance systems (Granberg et al., 2019). Termeer et al. (2017) advocate forms of continuous transformational change that relies on governance interventions focusing on generating small wins, amplifying these and overcoming institutional stagnation. Small changes can thereby amplify and accumulate into transformation of institutions, particularly in complex societal systems (Wagenaar, 2011). Continuous change, therefore, offers a fluid understanding of how agency and empowerment can develop in ways that facilitates an equitable mitigation of climate risks.

REFERENCES

Adger, W. N., et al. (2013). Changing social contracts in climate-change adaptation. *Nature Climate Change, 3*(4), 330–333. https://doi.org/10.1038/nclimate1751

Adger, W. N., et al. (2019). The social contract for climate risks: Private and public responses. In T. Scavenius & S. Rayner (Eds.), *Institutional capacity for climate change response: A new apporach to climate politics* (Paperback ed.). Roultedge.

Agarwal, A., et al. (2012). Climate policy processes, local institutions, and adaptation actions: Mechanisms of translation and influence. *WIREs Climate Change, 3*(6), 565–579. https://doi.org/10.1002/wcc.193

Barry, B. (2005). *Why social justice matter*. Cambridge University Press.

Begg, C. (2018). Power, responsibility and justice: A review of local stakeholder participation in european flood risk management. *Local Environment, 23*(4), 383–397. https://doi.org/10.1080/13549839.2017.1422119

Blaikie, P., & Brookfield, H. (1987). *Land degradation and society*. Methuen.

Bosomworth, K. (2018). A discursive-institutional perspective on transformative governance: A case from a fire management policy sector. *Environmental Policy and Governance, 28*(6), 415–425. https://doi.org/10.1002/eet.1806

Bosomworth, K., et al. (2017). What's the problem in adaptation pathways planning? The potential of a diagnostic problem-structuring approach. *Environmental Science & Policy, 76*, 23–28. https://doi.org/10.1016/j.envsci.2017.06.007

Bullard, R. D. (1990). *Dumping in Dixie: Race, class and environmental quality*. Westview.

Carr, E. R., & Nalau, J. (2023). Adaptation rationales and benefits: A foundation for understanding adaptation impact. *Climate Risk Management, 39*, 100479. https://doi.org/10.1016/j.crm.2023.100479

CRJ. (1987). *Toxic waste and race in the United States: A national report on racial and socio-economic characteristics of communities with hazardous waste sites.* Commission for Racial Justice (CRJ).

Demetriades, J., & Esplen, E. (2009). The gender dimensions of poverty and climate change adaptation. In R. Mearns & A. Norton (Eds.), *The social dimensions of climate change* (pp. 133–141). World Bank. https://doi.org/10.1596/978-0-8213-7887-8

Diffenbaugh, N. S., & Burke, M. (2019). Global warming has increased global economic inequality. *Proceedings of the National Academy of Sciences, 116*(20), 9808–9813. https://doi.org/10.1073/pnas.1816020116

Eriksen, S. H., et al. (2015). Reframing adaptation: The political nature of climate change adaptation. *Global Environmental Change, 35*, 523–533. https://doi.org/10.1016/j.gloenvcha.2015.09.014

Fünfgeld, A. (2019). Just energy? Structures of energy (in)justice and the Indonesian coal sector. In T. Jafry (Ed.), *Routledge handbook of climate justice* (pp. 222–236). Routledge.

GAO. (1984). *Siting of hazardous waste landfills and their correlation with racial and economic status of surrounding communities.* U.S. General Accounting Office (GAO).

Glicken, J. (2000). Getting stakeholder participation 'right': A discussion of participatory processes and possible pitfalls. *Environmental Science & Policy, 3*(6), 305–310. https://doi.org/10.1016/S1462-9011(00)00105-2

Glover, L., & Granberg, M. (2020). *The politics of adapting to climate change.* Palgrave Macmillan.

Goldman, M. J., et al. (2018). A critical political ecology of human dimensions of climate change: Epistemology, ontology, and ethics. *Wiley Interdisciplinary Reviews: Climate Change, 9*(4), e526.

Gough, I. (2010). Economic crisis, climate change and the future of welfare states. *Twenty-First Century Society, 5*(1), 51–64. https://doi.org/10.1080/17450140903484049

Granberg, M., et al. (2019). Can regional-scale governance and planning support transformative adaptation? A study of two places. *Sustainability, 11*(24), 6978.

Harvey, D. (2009). *Social justice and the city.* University of Georgia Press.

Hayward, B., & O'Brien, K. (2010). Social contracts in a changing climate: Security of what and for whom? In A. L. St. Clair, et al. (Eds.), *Climate change, ethics and human security* (pp. 199–214). Cambridge University Press.

Howarth, R. B., et al. (2013). Intergenerational justice. In J. Dryzek (Ed.), *The Oxford handbook of climate change and society* (pp. 338–352). Oxford University Press.

Hügel, S., & Davies, A. R. (2020). Public participation, engagement, and climate change adaptation: A review of the research literature. *Wiley Interdisciplinary Reviews: Climate Change, 11*(4), 1–20.

ILO. (2019). *Indigenous peoples and climate change: Emerging research on traditional knowledge and livelihoods.*

IPCC. (2014). *Climate change 2014: Impacts, adaptations and vulnerability.* Cambridge University Press.

IPCC. (2023). *Synthesis report of the sixth assessment report (AR6).* Intergovernmental Panel on Climate Change (IPCC).

Islam, N., & Winkel, J. (2017). *Climate change and social inequality.* UN Department of Economic and Social Affairs (DESA).

Jafry, T. (Ed.) (2019). *Routledge handbook of climate justice.* Routledge.

Jensen, A., et al. (2020). Climate policy in a fragmented world—Transformative governance interactions at multiple levels. *Sustainability, 12*(23), 10017. https://doi.org/10.3390/su122310017

Johansson, H., et al. (2016). Climate change and the welfare state: Do we see a new generation of social risks emerging? In M. Koch & O. Mont (Eds.), *Sustainability and the political economy of welfare* (pp. 94–108). Routledge.

Kasperson, R. E., & Kasperson, J. X. (2005). Climate change, vulnerability and social justice. In J. X. Kasperson & R. E. Kasperson (Eds.), *The social contours of risk* (Vol. I, pp. 301–321). Earthscan.

Kaswan, A. (2021). Distributive environmental justice. In B. Coolsaet (Ed.), *Environmental justice: Key issues* (pp. 21–36). Earthscan.

Klepp, S., & Chavez-Rodriguez, L. (2020). Governing climate change: The power of adaptation discourses, policies, and practices. In S. Klepp & L. Chavez-Rodriguez (Eds.), *A critical approach to climate change adaptation: Discourses, policies, and practices* (Paperback ed., pp. 3–34). Routledge.

Krauss, C. (1994). Women of color on the front line. In R. D. Bullard (Ed.), *Unequal protection: Environmental justice and communities of color.* Sierra Club Books.

Krauss, C. (1998). Challenging power: Toxic waste protests and the politicization of white, working-class women. In N. Naples (Ed.), *Community activism and feminist politics.* Routledge.

Lasswell, H. D. (1936). *Politics: Who gets what, when, how.* Whittlesey House.

Levy, B. S., & Patz, J. A. (2015). Climate change, human rights, and social justice. *Annals of Global Health, 81*(3), 310–322. https://doi.org/10.1016/j.aogh.2015.08.008

Malloy, J. T., & Ashcraft, C. M. (2020). A framework for implementing socially just climate adaptation. *Climatic Change, 160*(1), 1–14. https://doi.org/10.1007/s10584-020-02705-6

Malone, E. L., & Engle, N. L. (2011). Evaluating regional vulnerability to climate change: Purposes and methods. *WIREs Climate Change, 2*(3), 462–474. https://doi.org/10.1002/wcc.116

Manuel-Navarrete, D., & Pelling, M. (2015). Subjectivity and the politics of transformation in response to development and environmental change. *Global*

Environmental Change, 35, 558–569. https://doi.org/10.1016/j.gloenvcha.
2015.08.012

Marion Suiseeya, K. R. (2021). Procedural justice matters: Power, representation, and partcipation in environmental governance. In B. Coolsaet (Ed.), *Environmental justice: Key issues* (pp. 37–51). Earthscan.

Meister, R. (2002). Human rights and the politics of victimhood. *Ethics & International Affairs, 16*(2), 91–108. https://doi.org/10.1111/j.1747-7093.
2002.tb00400.x

Mendelsohn, R., et al. (2006). The distributional impact of climate change on rich and poor countries. *Environment and Development Economics, 11*(2), 159–178. https://doi.org/10.1017/S1355770X05002755

Meyer, L. H. (2016). *Intergenerational justice.* Routledge.

Miller, D. (1999). Social justice and environmental goods. In A. Dobson (Ed.), *Fairness and futurity: Essays on envronmental sustainability and social justice* (pp. 152–172). Oxford University Press.

Moloney, S., et al. (2018). Climate change responses from the global to the local scale: An overview. In S. Moloney, et al. (Eds.), *Local action on climate change: Opportunities and constraints* (pp. 1–16). Routledge.

Murdock, E. G. (2021). A history of environmental justice: Foundations, narratives, and perspectives. In B. Coolsaet (Ed.), *Environmental justice: Key issues* (pp. 6–17). Earthscan.

Nightingale, A. J. (2011). Bounding difference: Intersectionality and the material production of gender, caste, class and environment in Nepal. *Geoforum, 42,* 153–162.

O'Brien, K., et al. (2007). Why different interpretations of vulnerability matter in climate change discourses. *Climate Policy, 7*(1), 73–88.

Page, E. (1999). Intergenerational justice and climate change. *Political Studies, 47*(1), 53–66. https://doi.org/10.1111/1467-9248.00187

Parvin, P., & Saunders, B. (2018). The ethics of political participation: Engagement and democracy in the 21st century. *Res Publica, 24*(1), 3–8. https://doi.org/10.1007/s11158-017-9389-7

Patterson, J. J., et al. (2018). Political feasibility of 1.5°C societal transformations: The role of social justice. *Current Opinion in Environmental Sustainability, 31,* 1–9. https://doi.org/10.1016/j.cosust.2017.11.002

Paulson, S., et al. (2003). Locating the political in political ecology: An introduction. *Human Organization, 62*(3), 205–217. http://www.jstor.org/stable/44127401

Pelling, M. (2011). *Adaptation to climate change: From resilience to transformation.* Routledge.

Pelling, M., et al. (2015). Adaptation and transformation. *Climatic Change, 133*(1), 113–127. https://doi.org/10.1007/s10584-014-1303-0

Rawls, J. (1971). *A theory of justice.* The Belknap Press.

Reed, M. S. (2008). Stakeholder participation for environmental management: A literature review. *Biological Conservation, 141*(10), 2417–2431. https://doi.org/10.1016/j.biocon.2008.07.014

Rickards, L. (2013). Transformation is adaptation. *Nature Climate Change, 3*(8), 690–690.

Robbins, P. (2012). *Political ecology: A critical introduction.* Wiley-Blackwell.

Roberts, J. (2020). Political ecology. In F. Stein, et al. (Eds.), *Cambridge encyclopedia of anthropology.* University of Cambridge. https://doi.org/10.29164/20polieco

Skillington, T. (2019). *Climate change and intergenerational justice.* Routledge.

Sovacool, B. K. (2018). Bamboo beating bandits: Conflict, inequality, and vulnerability in the political ecology of climate change adaptation in Bangladesh. *World Development, 102*, 183–194. https://doi.org/10.1016/j.worlddev.2017.10.014

Sovacool, B. K. (2021). Who are the victims of low-carbon transitions? Towards a political ecology of climate change mitigation. *Energy Research & Social Science, 73*, 101916. https://doi.org/10.1016/j.erss.2021.101916

Sovacool, B. K., & Linnér, B.-O. (2016). *The political economy of climate change adaptation.* Palgrave Macmillan.

Sultana, F. (2022). Critical climate justice. *The Geographical Journal, 188*(1), 118–124.

Sze, J. (2021). Sustainability and environmental justice: Parallell tracks or at the crossroads. In B. Coolsaet (Ed.), *Environmental justice: Key issues* (pp. 107–117). Earthscan.

Taylor, M. (2015). *The political ecology of climate change adaptation: Livelihoods, agrarian change and the conflicts of development.* Earthscan.

Termeer, C. J. A. M., et al. (2017). Transformational change: Governance interventions for climate change adaptation from a continuous change perspective. *Journal of Environmental Policy & Planning, 60*(4), 558–576. https://doi.org/10.1080/09640568.2016.1168288

Terry, G. (2009). No climate justice without gender justice: An overview of the issues. *Gender & Development, 17*(1), 5–18.

Tokar, B. (2019). On the evolution and continuing development of the climate justice movement. In T. Jafry (Ed.), *Routledge handbook of climate justice* (pp. 13–25). Routledge.

Tol, R. S. J., et al. (2004). Distributional aspects of climate change impacts. *Global Environmental Change, 14*(3), 259–272. https://doi.org/10.1016/j.gloenvcha.2004.04.007

Vaccaro, I., et al. (2013). Political ecology and conservation policies: Some theoretical genealogies. *Journal of Political Ecology, 20*(1), 255–272. https://doi.org/10.2458/v20i1.21748

Wagenaar, H. (2011). *Meaning in action: Interpretation and dialogue in policy analysis*. M.E. Sharpe.

Walker, G. (2012). *Envorinmental justice: Concepts, evidence and politics*. Routledge.

Watts, M. (2000). Political ecology. In E. Sheppard & T. J. Barnes (Eds.), *A companion to economic geography* (pp. 257–274).

WCED. (1987). *Our common future*. Oxford University Press.

PART IV

Contrasting Climate Futures: Revealing Threats, Reshaping Values

CHAPTER 8

Conclusions on Climate Change as Societal Risk

Abstract In concluding the volume, the chapter reconnects to the book's aims, draws together the work's major themes and offers some overarching conclusions. After overviewing salient features of climate change as societal risk, the case for being concerned over societal risk is presented, drawing a distinction between general social risks and the special case of societal risk. Undertaking this inquiry dictates finding a suitable form of knowledge that extends beyond scientific-technical forms. Central to the chapter is the framework of societal risk that incorporates hazards, exposures and societal expressions of the resultant societal risks. Key social implications of the societal risk concept are reviewed. Closing the chapter are a discussion of the limits of the societal risk concept and the case for taking up the concept in policies, plans and programs.

Keywords Climate change · Socio-ecological systems · Societal risk · Societal risk framework

This volume explores the societal risks of climate change according to its stated aims (see Chapter 1). Accordingly, in the preceding chapters, we have identified the distinctive characteristics of climate change risks (Aim 1), reviewed the understanding of the climate–society nexus in this context (Aim 2), described and reviewed the major interpretations

© The Author(s), under exclusive license to Springer Nature
Switzerland AG 2023
M. Granberg and L. Glover, *Climate Change as Societal Risk*,
https://doi.org/10.1007/978-3-031-43961-2_8

of risk, covering scientific-technical and cultural-social risk theories (Aim 3) and identified and defined the societal risks of climate change (Aim 4). These findings are accumulated and synthesised in this concluding chapter, where the contrasting futures for the climate–society nexus are considered in the form of the implications of societal risk (Aim 5). In this final chapter, we revisit these aims.

Climate change risk has an elusive identity. It sits somewhat awkwardly in risk typologies dividing risk into either natural sources or those resulting from human activity (see the different contributions in, Downing et al., 1999). Because of climate change's anthropogenic character, it is neither entirely a conventional natural nor a technological hazard. Furthermore, climate change risk is difficult to circumscribe with a singular identity, as it comprises many different sorts of risks to social and ecological systems. Encompassing multiple categories of risks and crises, climate change can be 'sudden' in its impacts, 'creeping' with a slow onset or 'chronic', meaning it is "...ongoing ... with no obvious solution" (McConnell, 2003, p. 395). Characterising climate change as societal risk as a 'wicked problem', in reference to the absence of a single resolution and its dynamic character, is tempting. But arguably, it is more difficult and confounding than a single problem, since it is a category of problems and with great uncertainty and disagreement over its identity, magnitude and significance.

Climate change can be considered a societal risk with great potential for harm as impacts can be amplified or attenuated throughout social and ecological systems through complex relations and interdependencies (cf., van Asselt & Renn, 2011). Societal risks are transboundary and characterised by complexity, interdependency, lags in perception and regulation, non-linearity, tipping points and the holistic threat of potential irreversible system collapse (Aven & Renn, 2020; Schweizer, 2021). This raises the fundamental challenge of how to understand such risks, as this forms basis for responsive decision-making.

Contemporary research and scholarship and the governance responses to these climate threats, however, concentrates on these risks in isolation and usually singularly, such as to communities, economic sectors, essential infrastructure and services and vulnerable locations (see, e.g., IPCC 2022). Lacking has been considering climate change risks at the societal scale, hence the promotion of the societal risk concept in this volume.

WHY BE CONCERNED OVER SOCIETAL RISK?

Social risk is a useful concept for identifying both general and specific climate change risks across society. But, as argued in this volume, there are good reasons for considering those risks posed by climate change to the fundamental social unit, that of society—as opposed to social risks to societies' members or individual sectors (e.g., business, civil and government). This argument is based on acknowledging the benefits and utility of the broad and flexible concept of social risks but finding it insufficient for the problem of risks at the societal level. As suggested above, the difference between the social and societal is far greater than a taxonomy of mere scale and specificity. By analogy, our focus on societal risk, rather than social risk, is akin to difference between the risks of a sinking ship and its passengers and the risks to its passengers alone.

Accordingly, our use of societal risk (as compared to social risk) intends to circumscribe the specific climate change risks to societies for the following reasons:

- It can be defined in relatively exact way,
- It provides an avenue for investigating concerns with the root causes of social vulnerabilities entailing society in its entirety,
- It facilitates recognising the interconnections between social and related systems that produce risk,
- It provides a setting for assessing the universal risks to a society, albeit that these are likely to be highly differentiated within societies, and
- It furnishes a means to incorporate protecting ecological and environmental values as social concerns.

FINDING SUITABLE KNOWLEDGE OF CLIMATE CHANGE AS SOCIETAL RISK

Above, the book describes how scientific rationalism governs the analytical approach to climate change risks in the early decades of the climate change discourse (cf., Coen, 2021), as it had for disaster risk and management since the inception of relevant public policy interventions. Under positivist models, the risk of a specific hazard is, in its rawest form, simply 'likelihood times impact'. From this, a risk response can be formulated; characterised as a (rational) linear model, the basic approach is that, following risk identification, response options are devised along with appropriate selection criteria, from which an optimal response is selected (aka, the 'classical' decision approach) (Harry, 2008).

For recognising societal climate change hazards, the positivist model (and its associated rational model of response) suffers from limitations in its climate risk estimations because of, *Firstly*, the expansive character of climate change (see Chapters 2 and 3) and *Secondly*, the subjective character of the societal risk (see Chapter 5). On the first problem, the positivist model is highly problematic in reducing climate change to a suitable set of estimations, including that the phenomenon evolves across broad temporal, spatial and socio-political scales (see Table 8.1). On the second problem, the positivist model cannot properly accommodate the social influences on risk (a minor example is that empirical risks rarely align with those as perceived by individuals or groups) (see Chapter 3). For these reasons, traditional risk approaches for creating knowledge have to be supplemented or supplanted by those more suited to dealing with the uncertainty of climate change, and the character of society, for the task of addressing societal risk.

Eventually, these approaches to climate change risk give way to approaches acknowledging the multiplicity of interests, values and views of different stakeholders in the climate change impacts realm, albeit while holding that scientific-technical knowledge remains the foundation for understanding risk and formulating responses. Over time, the dominance of this realist model is challenged by constructivist approaches. Constructivist views admit other types of knowledge and perspectives, recognise a multiplicity of knowledges understood subjectively, interpret material realities through social relations and seek to identify the interests and values of knowledge types and other differences (see Chapter 4).

As described above, societal risk is a function of politics. Political actors, ideologies, institutions, interests, representations, resources and other aspects play a major role in distributing and shaping societal risks. Overlain on existing and historical circumstances, politics influences risk exposures, coping capacities, sensitivities to hazards and adaptive vulnerabilities. Societal transformation as the avenue for social change to fulfil the goals of climate justice (see Chapter 7), necessitates continued political campaigns for such a goal.

Table 8.1   Critique of traditional risk-based approaches to climate change

Technical critique of risk-based approaches	• False assumption of a single polity, • False assumption that impacts costs are a small proportion of resources available to address them, • False assumption that values are known and exogenously determined, • Rejection of economic discounting of future values, and • Rejection of climate science's assumptions of linear change (given that climate change can be non-linear)
Conceptual critique of risk-based approaches	• Risk rationality has falsely presumed that global hazards can be effectively measured, calculated and controlled, • Risk rationalities are used to express and reinforce political power, and • Scientific-technical risk culture is limited by its realist perspective

Source After, Morgan et al. (1999) and Pidgeon and Butler (2009)

Elaborated in this book's introduction, our understanding of societal risk is underpinned by a view of society, wherein societal risks are those circumstances where *constituent* elements of society/societies are threatened with degradation or loss (see Chapter 1). Accordingly, a societal risk is not simply a quantitative aggregation of the climate risks to individual persons, corporations, social groups or organisations. Rather, the climate risks of concern relates to threats to those features of society that underpin, facilitate and reinforce the existence, flourishing and growth of social relations. These elements include those of culture, distribution of resources and services, identity, governance, social functions, social structures and relations within society and with other societies. Losses to individuals, groups, social institutions or corporations, therefore, are not necessarily societal losses if the elements that form and maintain a society are unharmed. In other words, climate change as *a societal risk is a risk to a society as a whole, rather than only a risk to any of its members.* To circumvent any suggestions that the members of a society are in any way being devalued by adopting a societal perspective, to reiterate arguments advanced in earlier chapters, this is categorically not the case, as the goal of climate justice demands that the interests of all members of society be treated equitably and fairly, together with the interests of future generations (see, e.g., Owen, 2021). Societal risk is socially constructed and interpreted from these elements.

A Framework of Societal Risk

Following its aims, this volume undertakes identifying and defining the societal risk of climate change, and to reviewing the climate–society nexus in this context (see Chapter 1). Accordingly, drawn together here are the relevant findings. Societal risk is an outcome of the sequential factors stemming from the generation of hazards in climate and related systems, the influence of which is determined by the risk exposure in the interacting set of socio-ecological systems, that gives rise in turn to the societal expression of risk through the functioning of the four key social systems in the political, economic, institutional and cultural realms. These relationships and sequential phenomena can be organised into a framework of societal risk (see Table 8.2).

As outlined above, no specific sociological model of society is endorsed here, but rather the analysis focuses on a set of societal attributes

Table 8.2 Framework of societal risk

Risk aspect	Locus of determinant	Risk determinant
Hazard Source	Climate and Related Systems	• Current and future climate system impacts • Climate-related system impacts
Exposure	Socio-ecological Systems	• Ecological responsibilities • Non-material social/cultural systems and outputs • Socio-ecological systems and outputs • Social/cultural entities and processes
Societal Expression	Social Subsystems	• Cultural • Economic • Institutional • Political

comprising core and essential social functions (such that their degradation or loss results in societal decline or collapse). In the societal risk framework, societies are exposed to risk through a climate–society nexus, expressed through the interactions between these societal subsystems:

- Societal order and basic organisation are a function of the political application of power through political processes and structures, i.e., *political subsystems*. Although the institutions, modes and practices used for undertaking these functions vary enormously between societies, they share the common need to produce binding collective decisions; in modern industrial societies, their formal expressions are primarily manifested in executive, judicial and legislative bodies,
- Societies require systems for allocating and distributing material resources and associated opportunities for resource harvesting and use; in modern societies, this is an economic function, but noting that economics is too narrow and clumsy a concept when considering traditional societal exchanges (where sharing the outputs of common pool resources is typical), i.e., *economic subsystems*. Resource allocation is often tied to modes of production and associated social stratification,
- Central to society are those institutions tying its disparate elements and social groups into a cohesive whole and enabling collective efforts, i.e., *institutional subsystems*. This is an essentially sociological use of the concept, wherein complex (self-reproducing) social forms (prominently as organisations) undertake specific tasks (relating to discourse, economics, politics, law and others functions) through the use of conventions, norms, rules and the like, and
- Finally, there is the influence of culture, that for convenience is described here as *cultural subsystems*. Culture provides societies with collective identity, playing a role in societal beliefs, conventions, communication, education, knowledge formation, rituals and values (in a philosophical sense). Although all these subsystems interact and overlap, culture obviously permeates a society to the greatest extent; our identities and perceptions are phenomena most likely to be tied to the culture of our societies.

FACING CLIMATE CHANGE AS SOCIETAL RISK

Clearly, current and future social and ecological losses are unavoidable. When facing climate change as societal risk, the issue of what are acceptable, tolerable or intolerable risks has to be considered (Dow et al., 2013; Fischhoff et al., 1981). *Acceptable risks* are those for which investing additional effort/investment for their mitigation is deemed unnecessary. *Tolerable risks* can be worthwhile mitigating for other benefits and where limited risk reduction measures can be considered. *Intolerable risks* exceed social norms and can persist even with societal action to mitigate risk and reduce vulnerabilities. Handling climate change as a societal risk, therefore, will always entail prioritisation based on social values of some sort (e.g., cultural or economic).

To date, researchers and scholars have assembled a vast array of materials describing and forecasting climate change impacts in the natural and social realms. Of particular relevance to climate change risk is the research on planetary limits, in which climate change is but one factor in describing the extent to which the earth's resources and ecological systems are being stressed. Two salient lessons can be drawn from this research. *Firstly*, these planetary limits are drivers of risk that may combine with, and therefore, magnify climate risks, and *Secondly*, by inference, there is a better understanding of the systems that underpin the functioning of the planet, and the risks society is imposing upon it, than there is of climate change as societal risk. In other words, there is a great deal of knowledge of climate risks to social values and to ecology but, as it stands, there is no consolidated body of knowledge of the risks climate change poses to societies.

Society's survival depends on the proper functioning of its socio-ecological relationships furnishing critical nutrients, exogenous energy, materials, waste assimilation and other essential supplies and services (Davidson, 2023; IPCC, 2023). An insight from human ecology is that these relationships are mediated at the social scale through institutions, and the social infrastructures that they create and maintain (Hawley, 1986). A part of this relationship is, in the widest sense, knowledge, which includes technology. Societal collapse in history and pre-history (see Chapter 6), has often been explained in terms of the failings of key ecological services, but invariably such failings were at least partially founded in the social realm (Middleton, 2017).

Effective functioning of social systems is also essential for societal survival from climate impacts, including extreme events and disasters. Additional to immediate material and service losses in climate disasters are ongoing losses. Climate change impacts can overwhelm the capabilities of social systems and services dealing with extreme events and emergencies, such as emergency services, but also other relevant systems such as education, elder care and societal planning, through such factors as capital losses, fatigue and mounting financial costs, i.e., a decline in adaptive capacity. This is especially the case where recovery after losses to climatic impacts is incomplete, so that previous resilience is diminished.

Although dramatic events garner greater public attention, the slow climatic changes are equally hazardous and global in their extent, albeit with greater diversity and differing spatial and temporal features. Changes in climatic variables, such as to their duration, frequency and/or intensity, may produce acute problems, but accumulating chronic problems can also lead to losses to, or failings of, social systems. Escalating stresses on socio-ecological and social systems can reach critical thresholds, precipitating dramatic social losses or societal collapses. Differences within and between societies demonstrate the dynamic and context-dependant character of societal risk.

Social uncertainties connected to climate change as societal risk generate new sources of social instability, through conflicts within and between social groups and societies (Koubi, 2019; Lövbrand & Mobjörk, 2021; Nordås & Gleditsch, 2015) making judgements complicated. This potential of conflict, interdependent and cascading social impacts adds further to the uncertainties connected to climate change's physical implications. As equity and inequity are closely related to conflict and vulnerability to hazards risk, risk analysis needs to encompass issues of equity, justice and power (Owen, 2021). Therefore, applying a risk-based approach to climate change adaptation and planning involves acknowledging variations in risk perceptions, differences and interactions between drivers, vulnerabilities, governance, policy, actors and capacity to handle impacts (Simpson et al., 2021). Accordingly, society's handling of climate change as a societal risk requires a broad perspective in the risk identification, assessment, management and governance components.

Therefore, climate change as a societal risk poses particular, enormously diverse and fundamental challenges to society (see, e.g., IPCC, 2023):

- *Firstly*, the risk is characterised by the great uncertainty of future climate change in terms of direction of change, geographical context, timing of changes and scale and extent. Some climate change risks are sudden and acute, others will involve a creeping escalation in potential severity and harmful consequences, while others are chronic,
- *Secondly*, societal risks can have a global dimension when risks in one place result from causal factors in a distant but systemically connected other location (and may involve social systems with widely differing constituent elements),
- *Thirdly*, the larger climate change risks are likely to be cascading risks, as climate systems are usually part of interconnected key socio-ecological systems,
- *Fourthly*, risk assessment and response capacities vary greatly within and between societies, so that although different societies can have similar responses to similar risks, they can also have widely differing responses and, and as a result, differing vulnerabilities,
- *Fifthly*, the social values at risk are of great variety and type across cultures, locations, regimes, socio-economic and socio-ecological themes. Overlaid on this is the diversity of scales of social units, and
- *Sixthly*, investigating and understanding societal risk needs to be undertaken within a constructivist mode; for although there is a role for scientific-technical input, this should be incorporated into social science inquiries into society.

ARE THERE LIMITS TO THE SOCIETAL RISK CONCEPT?

Set against the claims presented here for promoting climate change as societal risk are some obvious limitations. Risk in the social realms (as with vulnerability and resilience) is perpetually dynamic and, in a sense, is always elusive and incomplete. Societal risk embodies the structural justices and injustices within society and expresses implications for future justice but is not an instrument for pursuing social justice—this must be sought through other means. Societal risk is, therefore, not ethically neutral. Knowing what is at societal risk holds a mirror to society, revealing what is valued. Societal risk places before society knowledge of what may be lost or harmed and is a reckoning of the extent to which society wishes to protect its assets, environments, institutions, people and services and the like. Risk assessment, however, invariably reflects the interests and values of those conducting the assessment. A societal risk framework can be (relatively) objectively formulated but its use can only be subjective. Furthermore, the orientation of risk assessment is invariably that of maintaining the *status quo*, but critically, can also be a foundation for processes of societal transformation.

These conditions give rise to several conundrums. *Firstly*, societal risk contains an inescapable anomaly by being an assessment of *current values* as threatened by *current and future risks*. *Secondly*, societal risk may include assessments of the very activities and processes that are giving rise to the risks to society, most prominently, the unsustainable use of local and global ecological materials and services. *Thirdly*, societal risks will invariably clash with the risks of some entities within society; it is Utopian to imagine that the interests of society and all its component entities and members are identical or aligned.

FINAL WORDS: WHY CONSIDER SOCIETAL RISKS FROM CLIMATE CHANGE?

In closing, a rationale and justification for considering climate change as societal risk is offered. How societies are organised and re-organised matters and is clearly connected to both the causes of, and vulnerabilities to, climate change (cf., Urry, 2011). Climate change risks need not be climate change impacts destiny. If societies are to be protected from the harms and losses from climate change impacts, then a systematic way of understanding the social context in which the risks are interpreted is required. Understanding politics and its implications is vital for interpreting the forms and features of societal risks. Politics is, additionally, central in defining and identifying key threats to society and in reshaping values in ways that contribute to positive transformation. While it is valuable to know the risks to people and ecology from climate change, it is also valuable to know the risks faced by societies as a whole. Perhaps the idiom 'You can't see the forest for the trees' encapsulates the problem of the myopia of overlooking the societal scale.

Knowledge of climate change as societal risk is essential in making decisions and formulating response plans, policies, practices and programs. Adaptation decisions cover such responses as preventing losses, tolerating losses, spreading or sharing losses, restoring lost assets and activities due to hazards, changing risk-exposed activities, shifting activities and communities and transforming climate–society relationships. Appreciating climate change as societal risk should shape adaptation and mitigation options, assessment criteria, preferences and priorities and resulting decisions. Such knowledge can also reveal the extent of existing capabilities and management abilities for adaptation. Considering climate change as societal risk can shift the concern about climate change risks from individual actors to entire societies, from actors in isolation to the relationships between such actors and from the symptoms of social risk to the root causes of societal risk.

No one can argue against knowing the climate change risks to individuals, households, businesses, economic sectors and institutions, but how useful is such knowledge without knowing the risks faced by the society that they inhabit? Perhaps the other dimension of understanding climate change as societal risk is the possibility to think more broadly and creatively about climate risks and to enlighten and provoke that most critical human faculty in responding to crises, our imaginations.

REFERENCES

Aven, T., & Renn, O. (2020). Some foundational issues related to risk governance and different types of risks. *Journal of Risk Research, 23*(9), 1121–1134. https://doi.org/10.1080/13669877.2019.1569099

Coen, D. R. (2021). A brief history of usable climate science. *Climatic Change, 167*(3), 51. https://doi.org/10.1007/s10584-021-03181-2

Davidson, J. P. L. (2023). Two cheers for collapse? On the uses and abuses of the societal collapse thesis for imagining anthropocene futures. *Environmental Politics.* https://doi.org/10.1080/09644016.2022.2164238

Dow, K., et al. (2013). Limits to adaptation to climate change: A risk approach. *Current Opinion in Environmental Sustainability, 5*(3), 384–391. https://doi.org/10.1016/j.cosust.2013.07.005

Downing, T. E., et al. (Eds.) (1999). *Climate change and risk.* Routledge.

Fischhoff, B., et al. (1981). *Acceptable risk.* Cambridge University Press.

Harry, C. (2008). Classical decision rules and adaptation to climate change. *Australian Journal of Agricultural and Resource Economics, 52*(4), 487–504. https://doi.org/10.1111/j.1467-8489.2008.00421.x

Hawley, A. H. (1986). *Human ecology.* The University of Chicago Press.

IPCC. (2022). Climate change 2022: Mitigation of climate change. Intergovernmental Panel on Climate Change (IPCC).

IPCC. (2023). *Synthesis report of the sixth assessment report (AR6).* Intergovernmental Panel on Climate Change (IPCC).

Koubi, V. (2019). Climate change and conflict. *Annual Review of Political Science*, *22*(1), 343–360. https://doi.org/10.1146/annurev-polisci-050317-070830

Lövbrand, E., & Mobjörk, M. (Eds.) (2021). *Anthropocene (in)securities: Reflections on collective survival 50 years after the Stockholm conference*. Oxford University Press.

McConnell, A. (2003). Overview: Crisis management, influences, responses and evaluation. *Parliamentary Affairs*, *56*(3), 363–409. https://doi.org/10.1093/parlij/gsg096

Middleton, G. D. (2017). *Understanding collapse: Ancient history and modern myths*. Cambridge University Press.

Morgan, M. G., et al. (1999). Why conventional tools for policy analysis are often inadequate for problems of global change. *Climatic Change*, *41*(3), 271–281. https://doi.org/10.1023/A:1005469411776

Nordås, R., & Gleditsch, N. P. (2015). Climate change and conflict. In S. Hartard & W. Liebert (Eds.), *Competition and conflicts on resource use* (pp. 21–38). Springer International Publishing. https://doi.org/10.1007/978-3-319-10954-1_3

Owen, G. (2021). Equity and justice as central components of climate change adaptation. *One Earth*, *4*(10), 1373–1374. https://doi.org/10.1016/j.oneear.2021.09.008

Pidgeon, N., & Butler, C. (2009). Risk analysis and climate change. *Environmental Politics*, *18*(5), 670–688. https://doi.org/10.1080/09644010903156976

Schweizer, P.-J. (2021). Systemic risks—Concepts and challenges for risk governance. *Journal of Risk Research*, *24*(1), 78–93. https://doi.org/10.1080/13669877.2019.1687574

Simpson, N. P., et al. (2021). A framework for complex climate change risk assessment. *One Earth*, *4*(4), 489–501. https://doi.org/10.1016/j.oneear.2021.03.005

Urry, J. (2011). *Climate change & society*. Polity.

van Asselt, M. B. A., & Renn, O. (2011). Risk governance. *Journal of Risk Research*, *14*(4), 431–449. https://doi.org/10.1080/13669877.2011.553730

Index